Leckie ✕ Leckie
Scotland's leading educational publishers

CfE Higher
BIOLOGY

SUCCESS GUIDE

CfE Higher **BIOLOGY** *SUCCESS GUIDE*

Angela Drummond

001/20022015

10 9 8 7 6 5 4 3 2 1

ISBN 9780007554386

Published by
Leckie & Leckie Ltd
An imprint of HarperCollins*Publishers*
Westerhill Road, Bishopbriggs, Glasgow, G64 2QT
T: 0844 576 8126   F: 0844 576 8131
leckieandleckie@harpercollins.co.uk   www.leckieandleckie.co.uk

Publisher: Peter Dennis
Project Manager: Craig Balfour

Special thanks to
Ink Tank (cover design)
Roda Morrison (copy-edit)
QBS (layout and illustration)
Louise Robb (proofreading)
Felicity Kendall (proofreading)

A CIP Catalogue record for this book is available from the British Library.

Acknowledgements

P3 James Watson and Francis Crick © A. BARRINGTON BROWN/ SCIENCE PHOTO LIBRARY; P19 Stem cell treatment © J.C. REVY, ISM/SCIENCE PHOTO LIBRARY; P30 Genomic sequencing © PHILIPPE PLAILLY/SCIENCE PHOTO LIBRARY; P39 Enzyme structure © J.C. REVY, ISM/SCIENCE PHOTO LIBRARY; P78 Tapeworm © ERIC GRAVE/SCIENCE PHOTO LIBRARY

'For all my pupils in Orkney, past and present.' Angela Drummond

Unit 1 – DNA and the Genome

Contents

Unit 2 – Metabolism and Survival

Contents

Unit 3 – Sustainability and Interdependence

Introduction

The CfE Higher

The CfE Higher Biology course offers relevant, up-to-date information on a broad range of topics in biological science, which provide the foundation for biological research today.

The course covers key principles of biology including molecular biology, biochemistry, physiology, genetics, plant science, zoology, biodiversity and agriculture. There is a great deal of exciting biology to learn within the course, and regular weekly revision of work is advisable in order to strengthen knowledge and understanding.

Structure

The course is SCQF level 6, and is worth 24 credit points.

The CfE Higher Biology Course is made up of three units:

- DNA and the genome
- Metabolism and survival
- Sustainability and interdependence

In order to achieve a pass at Higher, you need to pass all of these units as well as the course assessment.

Course assessment

The Course assessment assesses both attainment and added value, resulting in the award of an overall grade, within the range grade A to grade D.

The assessment takes the form of a question paper (exam) and a course assignment.

The exam

The exam gives you the opportunity to apply skills in both problem-solving and knowledge and understanding, answering questions from across the three units of the course.

The exam carries a total of 100 marks and is made up of two sections:

- the objective test (20 marks) comprising 20 multiple-choice questions
- Paper 2 (80 marks) which is made up of a mixture of restricted and extended response questions.

The assignment

The course assignment requires you to complete a challenging task in a topic selected from the three units of the course. There are two stages to the assignment, a research stage and a communication stage. The communication stage consists of a structured report of your findings, written under exam conditions, using only your research notes. It is worth 20 marks, which are added to your final exam mark, giving a total course mark out of 120.

The assignment gives you the opportunity to apply your skills of scientific inquiry through research.

Preparation for both of these elements of the course assessment is critical to getting the best grade possible in your Higher Biology course.

How to use this book

This Success Guide is designed to help you in both your revision work for Higher Biology, and to support you with current learning. The topics are broken down into concise sections that will assist you with recall and help you to lock in key points. Top tips in each chapter guide you towards success and help you to remember key facts. Quick tests at the end of each chapter are designed to ensure that you have learned essential facts in order to improve your performance in exam questions that test knowledge and understanding. Use the quick tests to regularly check that you've retained the important information on each topic.

Answers to quick tests can be found at the back of the book.

Throughout the book, examples are given, where possible, to illustrate biological concepts. Use of examples is an important element for success in extended response questions.

The guide follows the unit structure of the course as detailed in the SQA National Course Specifcation Document. It's advisable to use this book alongside other resources: it is not a one-stop route to exam success but should be an important tool in building up your knowledge and understanding of the course and your confidence in your abilities to apply that learning.

Success in Higher Biology also depends on your knowledge of new vocabulary. Many assessment questions test your knowledge of the meaning of a particular word or phrase. To help you learn words and meanings, a glossary of terms has been included at the back of this book.

Use this book as a starting point – come back to it for the essential knowledge and skills that you need to tackle questions and understand concepts and processes in biology. This book may also provide useful support during biology lessons, and can be used as a quick reference book in addition to other resources.

Structure of DNA

Genetic information within every living cell is contained within **deoxyribonucleic acid** or DNA. It is a double stranded ladder shaped molecule, twisted in one direction to form a **double helix** shape.

Nucleotide structure

Subunits called **nucleotides** join together to form the two sides of the DNA molecule. A nucleotide consists of a phosphate group, deoxyribose sugar and a nitrogenous base. The deoxyribose sugar molecule has five carbon atoms. Carbon atom 1 is on the right, then count anti-clockwise to carbon atoms 2, 3 and 4, with carbon atom 5 situated between carbon atom 4 position and the phosphate group.

Direction of deoxyribose sugar and phosphate bonding

TOP TIP

Remember: free single nucleotides can join on to the DNA molecule at the 3' ends only.

Alternate phosphate groups and deoxyribose sugar molecules make up the two strands or sides of the 'ladder' shape, sometimes referred to as the 'sugar phosphate' backbone. They connect in opposite directions within each of the two strands, meaning they are **antiparallel**. Within one strand, each phosphate group forms a strong chemical bond with carbon atom 3 of the deoxyribose sugar molecules forming a chain that begins with a carbon atom 3 (3'), and ends with a carbon atom 5 (5'). The opposite strand begins with a 5' end and finishes with a 3' end.

A strong chemical bond forms between alternate phosphate and sugar molecules.

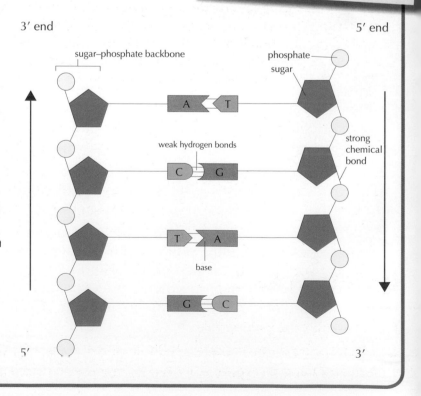

DNA base pairing

DNA has four nitrogenous bases which combine in pairs to form the 'rungs' of the ladder shaped molecule. Pairing always follows the same rule: adenine pairs with thymine, and cytosine pairs with guanine. Base pairs are held in position within the DNA molecule by weak hydrogen bonds.

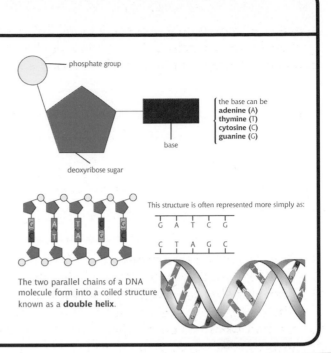

phosphate group

the base can be
adenine (A)
thymine (T)
cytosine (C)
guanine (G)

base

deoxyribose sugar

This structure is often represented more simply as:

G A T C G

C T A G C

The two parallel chains of a DNA molecule form into a coiled structure known as a **double helix**.

Discovering the structure of DNA

In the 1940s, a research scientist called Erwin Chargaff discovered that the ratio of thymine bases in DNA was equal to the ratio of adenine bases. The ratio of guanine bases was found to equal that of cytosine bases.

Using X-ray diffraction patterns of DNA produced by scientists Wilkins and Franklin, Watson and Crick worked out the shape of the DNA molecule. They built a model of the coiled ladder shape called a double helix, for which they won the Nobel prize in 1953.

James Watson and Francis Crick

Quick Test 1

1. State the base pairing rule.
2. What is meant by the description '5' and 3' ends' of a single DNA strand?
3. Why is DNA described as having an 'antiparallel' structure?

Organisation of DNA

In both prokaryotic and eukaryotic cells, DNA is organised into structures called chromosomes. In **prokaryotes** (such as bacteria) chromosomes are circular. In **eukaryotes** (such as animal and plant cells) they are linear.

Prokaryotic cells

A prokaryotic cell does not have a membrane bound nucleus, for example, a bacterial cell. DNA is found in the cytoplasm in the form of a large circular chromosome. Many smaller rings of DNA called **plasmids** are also present in the cytoplasm.

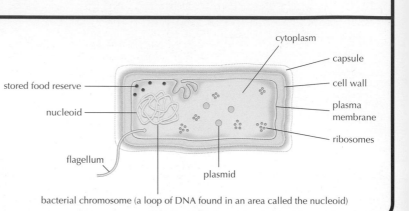

Eukaryotic cells

A eukaryotic cell has a membrane bound nucleus containing a number of linear chromosomes, for example muscle cells or cells from a plant. The DNA molecule that makes up a chromosome can be several metres long, and so is very tightly coiled around supporting bundles of **histone** proteins to form **nucleosomes**. Nucleosomes wrap together to form **chromatin** fibres, which loop, fold and condense into a **chromosome**.

Small circular rings of DNA are found in the **chloroplasts** and **mitochondria** of plant cells, and the mitochondria of animal cells.

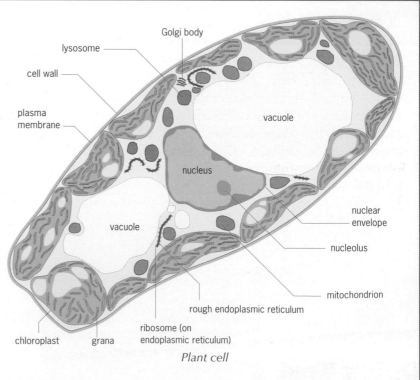

Plant cell

Chromosomes condense in eukaryotic cells during cell division, when they become visible with a microscope.

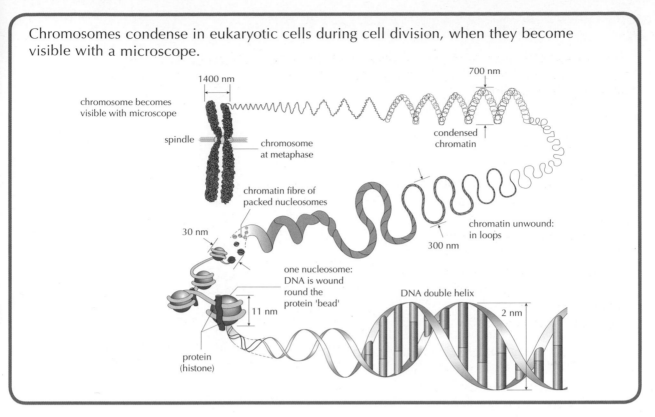

Yeast cells

Yeast cells are eukaryotic plant cells, and are a type of single celled fungus. They contain linear chromosomes within a membrane bound nucleus. They are unusual as they also contain some circular **DNA plasmids** within their cytoplasm, similar to those found in prokaryotes.

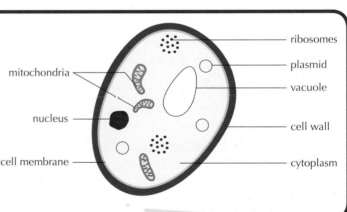

TOP TIP

Bacterial plasmids are used in genetic engineering. They are extracted from bacteria, cut open with **endonuclease** enzyme leaving **'sticky ends'**. A gene from another organism is placed between the 'sticky ends' and the plasmid is sealed with **ligase** enzyme. The bacterial cell now has a gene that can code for a foreign protein.

TOP TIP

Cell measurements:
1 mm = 1000 micrometers (μ)
1 μ = 1000 nanometers (n)

Quick Test 2

1.	What is the difference between a prokaryote and a eukaryote?
2.	Name the small ring of DNA in a prokaryotic cell.
3.	Why are yeast cells unusual in their organisation of DNA?

Replication of DNA

Copying the DNA code

The DNA in the nucleus of a cell must make an exact copy of all genetic information (replicate) before a cell can divide by mitosis. The DNA molecule acts as a template for DNA replication.

The DNA molecule untwists, then weak hydrogen bonds between the bases break using energy from ATP. The two **parental strands** then begin to separate or 'unzip', forming a Y shaped **replication fork**.

The formation of the first new **daughter strand** of DNA begins with the formation of a **primer**. This is a short sequence of DNA nucleotides formed at the 3′ end of a parental strand of DNA, furthest from the replication fork. The bases of the primer attach to complementary bases on the parental strand according to the base pairing rule.

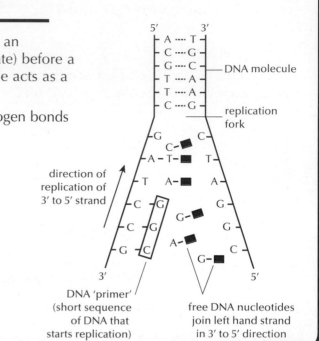

direction of replication of 3′ to 5′ strand

DNA molecule

replication fork

DNA 'primer' (short sequence of DNA that starts replication)

free DNA nucleotides join left hand strand in 3′ to 5′ direction

Process of DNA replication

The enzyme DNA polymerase adds free DNA nucleotides from the nuclear cytoplasm to their complementary base partners on the parental strand, beginning at the 3′ end of the primer. DNA polymerase also forms strong chemical bonds between the phosphates and sugars of the new DNA strand. Weak hydrogen bonds form between the base pairs of the new strand and the parental strand.

The formation of the second daughter strand begins with the primer attaching to the opposite parental strand that has the 5′ end, just below the replication fork. Free DNA nucleotides attach to the 3′ end of the primer in short lengths called **fragments**. The formation of each new fragment must begin with a primer. DNA polymerase forms strong chemical bonds between the phosphates and sugars of the fragments. When a fragment is complete, the primer is replaced by DNA. The enzyme ligase joins all of the fragments together to form the second new daughter strand.

TOP TIP

DNA polymerase can only **add nucleotides to a primer**, not directly to the beginning of the DNA strand.

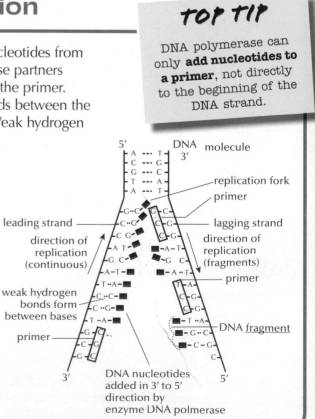

DNA molecule

replication fork

primer

leading strand

lagging strand

direction of replication (continuous)

direction of replication (fragments)

primer

weak hydrogen bonds form between bases

DNA <u>fragment</u>

primer

DNA nucleotides added in 3′ to 5′ direction by enzyme DNA polmerase

Leading strand

Replication of DNA from the 3′ end is **continuous** moving **towards** the junction of the replication fork; this is called the **leading strand**.

Lagging strand

Replication from the 5′ end is **discontinuous** moving **away** from the replication fork; this is called the **lagging strand**.

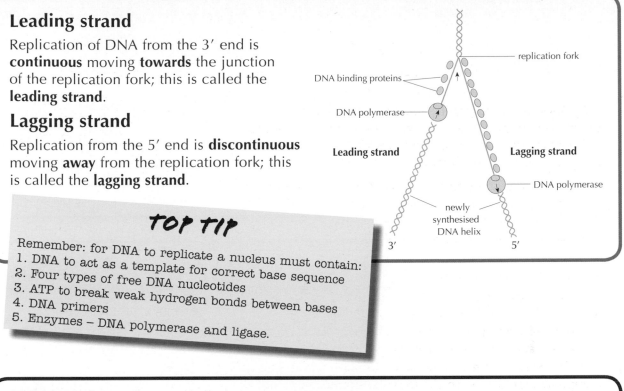

TOP TIP

Remember: for DNA to replicate a nucleus must contain:
1. DNA to act as a template for correct base sequence
2. Four types of free DNA nucleotides
3. ATP to break weak hydrogen bonds between bases
4. DNA primers
5. Enzymes – DNA polymerase and ligase.

DNA polymerase and DNA ligase

DNA polymerase is a multienzyme complex which joins free DNA nucleotides onto an exposed single strand forming a complimentary strand. The presence of a primer stimulates the action of DNA polymerase.

DNA ligase joins phosphates and sugars together which have been 'missed' by DNA polymerase. The action of both enzymes in DNA replication is similar.

Quick Test 3

1.	What is the function of a 'primer'?
2.	Free DNA nucleotides join on to which end of a primer?
3.	Which enzyme joins DNA fragments together?
4.	Why is replication from the 5′ end of the DNA molecule called 'discontinuous'?

Polymerase chain reaction (PCR)

Tiny fragments of DNA may be copied to provide enough for analysis. This is called **amplification** of DNA, and is done using a process called the **polymerase chain reaction (PCR)**. One cycle of PCR involves three steps carried out at three different temperatures, a process called **thermocycling**.

Primers

In the laboratory, a single strand of DNA is made with a complementary base sequence to the beginning of the DNA fragment to be amplified. This artificial strand of DNA is called a primer and is used to locate the specific DNA target sequence to be copied.

The process of PCR

1. The DNA fragment to be copied is mixed with a pool of free DNA nucleotides and heat resistant DNA polymerase enzyme.

   The DNA polymerase used in PCR is extracted from bacteria *Thermophilus aquaticus* (Taq) which live on the edge of volcanoes. This Taq polymerase is **not denatured by high temperatures**.

2. Two primers are made in the laboratory which have matching base pairs to the 3′ ends of the DNA fragment and are added to the mixture.

3. The mixture is heated to 93°C, then 55°C, 72°C then back to 93°C. This is one complete cycle.

   - 93°C – Two DNA strands separate
   - 55°C – Primers added, bind to DNA strands
   - 72°C – DNA polymerase adds DNA nucleotides to separate strands starting at primers. Two new double stranded copies of original DNA fragment are made.

The DNA fragment to be copied goes through 30 cycles per hour, making one million copies of the original fragment for analysis!

piece of DNA to be amplified

Heat to 93°C: the two strands separate

Add the primers and cool to 55°C so that they bind to the DNA

Raise temperature to 72°C. The thermostable polymerase enzyme copies each strand, starting at the primers

enzyme

enzyme

Repeat the process until enough DNA is made

Uses of PCR

DNA fragments can be amplified and used:

1. For DNA fingerprinting, to identify a person by blood or tissue left at a crime scene. Genes on the crime scene fragment are compared with a similar fragment of DNA obtained from suspects. If the genes match, a positive identification may be made.

2. To test for genetic diseases in a growing embryo, using embryonic cells.

3. Using chloroplast DNA to research plant evolution, analysis can provide information about the evolutionary relationships between plants, and allows plants to be classified according to similarities and differences in their DNA.

4. PCR can be used to amplify DNA fragments from two possible fathers and a child. This technique is used to confirm the paternity of a child.

DNA analysis using PCR

When copies of a DNA fragment have been made, some may be treated with restriction enzymes that cut the fragments into smaller pieces at specific base sequences. These fragments are separated out by gel electrophoresis to form a pattern of bands. This pattern is called a DNA fingerprint and is used to compare samples. If two bonding patterns are identical, then the DNA samples have came from the same person.

TOP TIP

Examination questions are likely to ask about the temperatures used in PCR, and what happens at each temperature.
- 93°C – Two strands of DNA separate.
- 55°C – Primers bind (**anneal**) to 3′ ends of fragment.
- 72°C – DNA polymerase adds free nucleotides until fragment is copied.

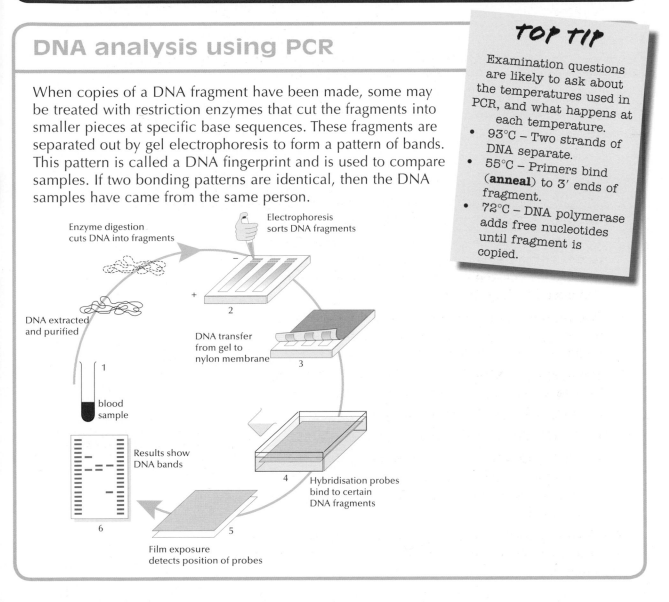

Enzyme digestion cuts DNA into fragments

Electrophoresis sorts DNA fragments

DNA extracted and purified

DNA transfer from gel to nylon membrane

blood sample

Results show DNA bands

Hybridisation probes bind to certain DNA fragments

Film exposure detects position of probes

Quick Test 4

1. Give two uses of PCR in everyday life.
2. Why is Taq polymerase used in PCR?
3. What is a 'primer' and why are they needed in PCR?
4. Which temperatures are used in one complete cycle of PCR?
5. What is meant by a DNA fingerprint?

Control of gene expression

Gene expression and phenotype

Proteins are made up of amino acid subunits. There are 20 different amino acids in the human body, which are absorbed from the products of digestion.

The **phenotype** or physical appearance of an organism is determined by proteins coded for by genes. **Gene expression** occurs when a gene is actively coding for a protein within a cell, regulated by the processes of **transcription** and **translation**. These processes turn the information from the DNA base sequence within a gene into the correct amino acid sequence required to form a specific protein. Only some genes within a cell are expressed at any one time.

Eukaryotic cells continually switch genes 'on' and 'off' in response to signals from the intra- and extracellular environments. If a gene is 'switched on' it is coding for a protein. Only some genes within a cell are switched on or 'expressed' at any one time.

Ribonucleic acid (RNA)

This is a second nucleic acid involved in gene expression or **protein synthesis**, which differs from DNA as follows:

- RNA is a single stranded molecule
- RNA has the base uracil instead of thymine (uracil pairs with adenine)
- RNA has a Ribose sugar
- RNA molecules much smaller than DNA.

Types of RNA

- **Messenger RNA** (mRNA) are formed in the nucleus from free RNA nucleotides which attach to base partners on an exposed gene. The strand of mRNA transports a copy of the base sequence of the gene to the cytoplasm of the cell, where it attaches to a **ribosome**. This is called **transcription**.

- **Transfer RNA** (tRNA) are molecules in cytoplasm which carry specific amino acid molecules to mRNA, placing them in the order determined by the base sequence of mRNA, to form a **polypeptide** chain. This is called **translation**.

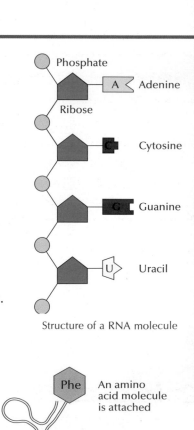

Phosphate

A ← Adenine

Ribose

C ← Cytosine

G ← Guanine

U → Uracil

Structure of a RNA molecule

Phe — An amino acid molecule is attached

▯ = Adenine

The mRNA passes through the ribosomes, and the tRNA brings together the amino acids

Structure of a tRNA molecule

- **Ribosomal RNA** (rRNA) form the structure of ribosomes together with proteins, which coordinate the process of translation.

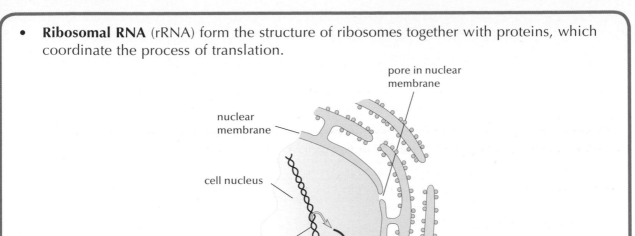

Introns and exons

Introns and **exons** are found in both DNA and RNA.

- Introns are regions which do not carry information about protein synthesis, often known as 'junk' or **non-coding** regions.
- Exons are regions of a nucleic acid molecule which are **active** in gene expression, and are known as **coding** regions.

Quick Test 5

1.	List the differences between DNA and RNA molecules.
2.	Identify the base partner of the RNA base uracil.
3.	What type of information is carried on an intron?
4.	Which form of RNA is involved in the translation stage of protein synthesis?
5.	How many different amino acids are there in the human body?

Transcription

The process of copying the base sequence of a gene within DNA to form a single strand of RNA occurs in the nucleus of a cell and is called transcription. DNA in the nucleus acts as a template for the production of a single strand of mRNA. DNA cannot leave the nucleus of a cell. It is the mRNA molecule that takes a copy of the base sequence of a gene from the nucleus out to the cytoplasm.

Stages of transcription

- A length of DNA (a gene) untwists and the two strands separate. Weak hydrogen bonds between the bases break.
- **RNA polymerase** assembles RNA nucleotides along **one** strand of the gene called the **sense strand**.
- A single strand of messenger RNA (mRNA) is formed from the RNA nucleotides, which carries a 'mirror image' of the base sequence of the gene on DNA. RNA polymerase forms strong chemical bonds between the ribose sugar and phosphate molecules.
- mRNA 'peels off' DNA and leaves the nucleus through a pore in the nuclear membrane. It attaches to a ribosome on the **rough endoplasmic reticulum (RER)** within the cytoplasm of the cell.

= Adenine ▮ = Guanine ⬠ = Uracil ⬡ = Thymine ▪ = Cytosine

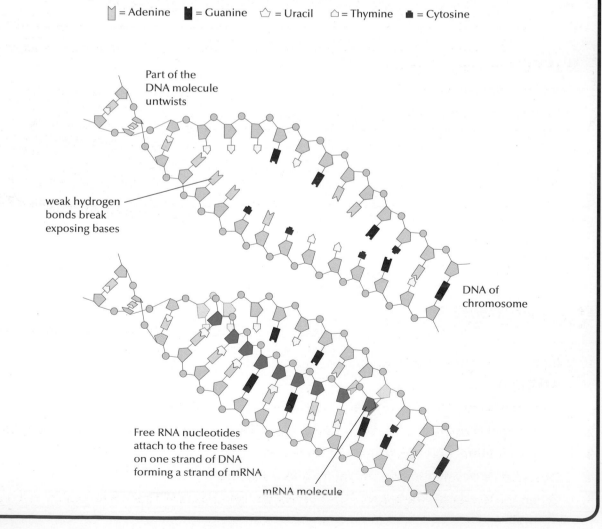

Part of the DNA molecule untwists

weak hydrogen bonds break exposing bases

DNA of chromosome

Free RNA nucleotides attach to the free bases on one strand of DNA forming a strand of mRNA

mRNA molecule

mRNA splicing

- mRNA contains both introns and exons and is called a **primary mRNA transcript**.
- Before leaving the nucleus, primary mRNA is cut by enzymes to remove the introns.
- Exons are then **spliced** (joined) together to form a **mature mRNA transcript**.
- The mature mRNA transcript then leaves the nucleus containing only exons, which can then be expressed as protein molecules.

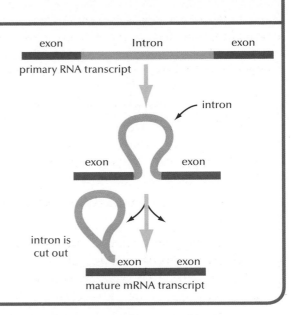

Maintaining the DNA base sequence

Doing transcription in RNA base sequence is a 'mirror image' of the DNA base sequence with the gene being expressed.

DNA base sequence – A T T C G G A T C

mRNA base sequence – U A A G C C U A G

TOP TIP

When a DNA base sequence is 'transcribed', it is changed to an RNA base sequence. The information contained **within** the base sequence is not changed at any time.

Quick Test 6

1. If part of the DNA base sequence on a gene is:

   AAG CGT GGT ATG ACC

   What would be the corresponding base sequence on mRNA?

2. Which strand of DNA is involved in the synthesis of a single strand of RNA nucleotides?

3. What is the function of RNA polymerase?

4. Describe the formation of a mature mRNA transcript.

5. When mRNA leaves the nucleus and enters the cytoplasm, it attaches to which cellular structure?

Translation

The process of assembly of amino acids in the correct order to form a polypeptide chain, according to the base sequence on mRNA, is called translation.

Ribosomes

Most ribosomes are found on the rough endoplasmic reticulum (RER) in a cell. There are two 'attachment sites' on the ribosome called site P and site A.

A small mobile subunit of a ribosome holds site E.

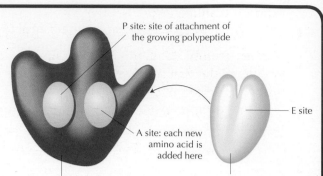

P site: site of attachment of the growing polypeptide

E site

A site: each new amino acid is added here

large subunit of the ribosome

small subunit of the ribosome

- Site P is the attachment site for the first tRNA molecule to join on to the mature mRNA transcript.
- Site A allows all following tRNA molecules to attach to the mRNA.
- Site E is located on a small subunit molecule, moves on to the ribosome and releases the completed polypeptide chain.

Stages of translation

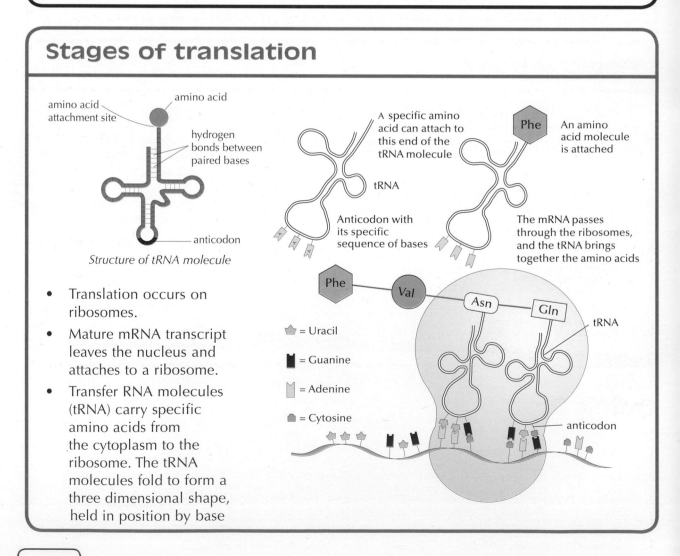

amino acid attachment site

amino acid

hydrogen bonds between paired bases

anticodon

Structure of tRNA molecule

A specific amino acid can attach to this end of the tRNA molecule

tRNA

Anticodon with its specific sequence of bases

Phe

An amino acid molecule is attached

The mRNA passes through the ribosomes, and the tRNA brings together the amino acids

Phe   Val   Asn   Gln

tRNA

= Uracil

= Guanine

= Adenine

= Cytosine

anticodon

- Translation occurs on ribosomes.
- Mature mRNA transcript leaves the nucleus and attaches to a ribosome.
- Transfer RNA molecules (tRNA) carry specific amino acids from the cytoplasm to the ribosome. The tRNA molecules fold to form a three dimensional shape, held in position by base

pairing. There are three bases at one end of their molecule called the **anticodon**, and an amino acid attachment site at the other end. The specific amino acid carried by a tRNA molecule depends upon the anticodon.

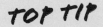

TOP TIP

Remember: Gene – DNA base sequence – mRNA codons – tRNA anticodons – amino acid sequence. Be able to identify a 'stop codon' and a 'start codon'.

- A group of three bases on mRNA is called a **codon**.
- A **start codon**, 'AUG' positioned on attachment site P, allows the first tRNA molecule carrying a specific amino acid to attach at the ribosome.
- The anticodons of the tRNA molecules attach to the codons of mRNA according to the base pairing rule, bringing amino acids into the correct position.
- **Peptide bonds** form between neighbouring amino acids and 'empty' tRNA molecules leave the ribosome and return to the cytoplasm to collect another specific amino acid molecule.
- There are three **stop codons** on mRNA: UGA, UAG and UAA. When a stop codon moves on to a ribosome at attachment site A, no tRNA molecules can now attach and the finished polypeptide chain is released.

TOP TIP

Polypeptide chains can be arranged to form different proteins such as enzymes, antibodies, hormones and structural proteins.

Peptide bond

A strong peptide bond forms between the carbon end of one amino acid and the nitrogen end of another through the removal of water (condensation reaction).

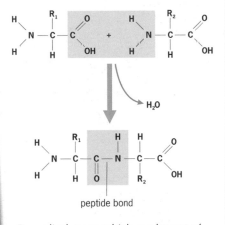

peptide bond

R = radical group which can be one of many interchangeable organic groups.

TOP TIP

On a large sheet of paper, draw a labelled diagram of the processes of transcription, translation and post-translational modifications. Use different coloured pencils to represent mRNA, tRNA, codons, anticodons and so on. A visual representation of these processes, and how they connect to one another, is often easier to understand and remember. Test the depth of your own knowledge by drawing out the 'story' of protein synthesis from memory!

Quick Test 7

1.	What is an 'anticodon'?
2.	How many different amino acids can a tRNA molecule carry?
3.	Which type of chemical bond forms between the amino acids on a ribosome?
4.	What is a 'stop codon'?
5.	Describe the function of binding sites 'P' and 'A' on a ribosome.

Structure of proteins

Protein composition

Proteins contain the elements carbon, hydrogen, oxygen and nitrogen. They are made up of amino acid subunits, linked by peptide bonds. A protein may contain up to 500 amino acids of 20 different types. Protein molecules have a 3D structure, where hydrogen bonds, electrostatic forces or sulfur atoms link the folds in polypeptide chains together. There are three main types of protein structure:

1. **Primary** – A single polypeptide chain of amino acids.

2. **Secondary** – A chain of amino acids folds upon itself, the resulting shape is held in place by hydrogen bonds.

3. **Tertiary** – A number of chains of amino acids fold around each other forming a complex protein molecule. Hydrogen, ionic and sulfur bonds hold the molecule in shape.

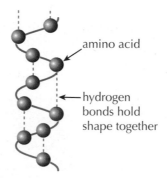

amino acid

← hydrogen bonds hold shape together

More than one protein can be expressed from a gene due to alternative RNA splicing, and post-translational modification.

Alternative RNA splicing

Different types of mRNA molecules can be produced from the same primary transcript, depending on which **parts** of the mRNA molecule are treated as exons, and which parts as introns. This is controlled by **regulatory proteins** specific to the cell type. In this way, **one gene can code for more than one protein**.

Post-translational modification

The arrangement and treatment of polypeptide chains following the process of translation determines the type of finished protein produced by the cell. Fibrous proteins such as collagen have polypeptide chains arranged in parallel rows, whereas globular proteins such as enzymes have polypeptide chains arranged in a three dimensional 'bundle' shape.

Post-translational protein structure may also be modified by the processes of molecular addition and cleavage.

Molecular addition

Some proteins must be combined with a 'non-protein' element before they can work properly. This is called molecular addition.

- Haemoglobin, which carries oxygen in blood, is a globular protein molecule combined with iron atoms. Haemoglobin is an example of a **conjugated protein**.
- Mucus is a globular protein molecule combined with polysaccharide (sugar) atoms. Mucus is an example of a **glycoprotein**.
- Egg yolk contains phosphoric acid combined with protein molecules.

Cleavage

Some polypeptide chains must be 'cut' by an enzyme for the protein to be 'activated' – this is called cleavage. For example, the hormone insulin only becomes active when the polypeptide chain is cut into two by enzymes following translation. The two separate parts of the polypeptide chain each fold over and their shapes are stabilised by sulfur bridges.

Haemoglobin – a conjugated protein

iron atoms

iron atoms

globular protein

TOP TIP

A **polypeptide** is a straight chain of amino acids joined by peptide bonds.
A **protein** is the completed molecule at the end of protein synthesis which is able to carry out a specific function within the body.

Quick Test 8

1.	Proteins contain which four chemical elements?
2.	Which type of bonds or forces hold 3D protein molecules in shape?
3.	Explain how one gene is able to code for more than one protein.
4.	Give one example of post-translational modification of protein molecules.

Differentiation in cells

A basic unspecialised cell becomes altered to have a specific structure and function through the expression of certain genes. This process is called differentiation.

Differentiation in plant cells

Undifferentiated cells dividing by mitosis are found in specific regions of a plant called **meristems**.

- **Apical meristem** is a region of mitosis in the root and shoot tip of a plant. Cell division results in vertical growth of the plant.

- **Lateral meristem** is a region of mitosis within a ring of **cambium** found in the stem of a perennial plant. Cell division results in an increase in the diameter of the stem.

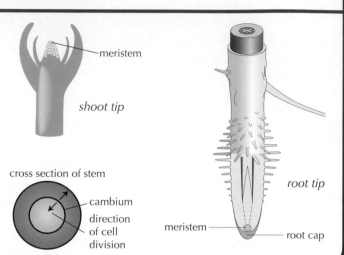

shoot tip

cross section of stem

root tip

meristem

cambium

direction of cell division

meristem

root cap

Differentiation in animal cells

Stem cells in the body have a basic structure and no specific function. They are able to:

- divide by mitosis to produce more stem cells, called **self renewal**

- differentiate into a specialised cell depending on gene expression.

Adult stem cells

Adult stem cells, found in the bone marrow, differentiate into blood cells only.

Adult skin stem cells differentiate into new skin cells only.

Embryonic stem cells

Embryonic stem cells have the potential to differentiate into any type of body cell, and so are valuable in research and in the treatment of human diseases.

The source of embryonic stem cells is a four day old fertilised human egg called a **blastocyst**.

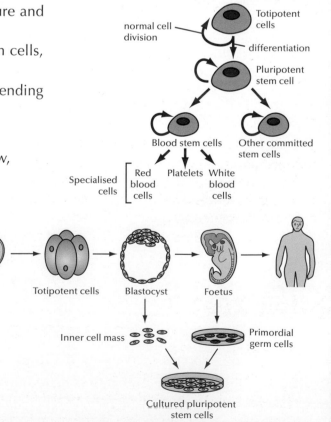

normal cell division

Totipotent cells

differentiation

Pluripotent stem cell

Blood stem cells

Other committed stem cells

Specialised cells

Red blood cells

Platelets

White blood cells

Totipotent cells

Blastocyst

Foetus

Inner cell mass

Primordial germ cells

Cultured pluripotent stem cells

Stem cell research

Research using stem cells can provide an understanding of how specific body cells will react to a new drug treatment or a specific disease. It can also provide a greater understanding of the process of cell division and cell differentiation within the body.

As a result of research, stem cells are now being used to treat a number of medical conditions in humans. Stem cell research is closely regulated by the government in Great Britain. Research teams wishing to use stem cells must first obtain a licence giving them permission to begin their research.

Stem cell treatments

- New skin can be grown in the laboratory from the stem cells of burns patients for grafting. As the new skin has been grown from the patient's own stem cells, it will not be rejected by the body.
- Bone marrow transplants give patients suffering from cancer of the blood (leukemia) the ability to make healthy blood cells from the stem cells within the donated bone marrow.
- Corneal replacement in the eye can be achieved by growing new corneal tissue from a patient's eye stem cells which replaces damaged cornea, restoring sight.

Ethical issues

There is debate about the destruction of human embryos as a means of supplying stem cells for research.

Induced pluripotent stem cells are now increasingly being used instead. These are adult cells which have had four specific genes added to them that allow them to differentiate into any type of body cell.

> ### TOP TIP
> Stem cells are **undifferentiated** which means they have a basic structure but no specific function.
> - **Totipotent** – A stem cell can differentiate into any type of body cell.
> - **Pluripotent** – A stem cell can differentiate into some types of body cell, but not all.

Quick Test 9

1. Describe a stem cell.
2. Name two sources of stem cells.
3. Why are embryonic stem cells so valuable in research?
4. Name two uses of stem cells in therapeutic medicine.
5. State the two locations of meristematic regions within a plant.

Structure of the genome

The genome

All the information within genes present in the DNA of a living organism is called the **genome**.

A section of DNA containing a base sequence which codes for a protein is called a **gene**.

The genome is made up of both coding sequences called **exons** and non-coding sequences called **introns**.

Function of 'non-protein coding sequences' or introns

- Regulation of transcription occurs when 'activator molecules' attach to non-protein coding sequences, which activate the RNA polymerase.
- Act as 'buffers' at the ends of chromosomes called **telomeres** which prevent chromosome damage.
- Code for the production of tRNA and ribosomal RNA.

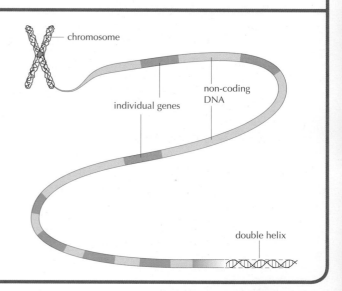

chromosome

individual genes

non-coding DNA

double helix

Mutation

A mutation is an irreversible change in the sequence of nucleotides within a gene or chromosome, resulting in either no protein or a defective protein being expressed.

A change in **genotype** due to a mutation causes a change in phenotype of the organism, which is referred to as a **mutant**.

Mutation is the only source of original genetic variation on which the process of **evolution** depends.

Mutagenic agents

The frequency of mutations may be increased by mutagenic agents such as gamma rays, ultraviolet light, cigarette smoke and mustard gas.

TOP TIP

Random mutation of the genetic material within a cell is a source of **genetic variation** which forms the basis of the processes of natural selection, speciation and evolution.

Single gene mutation

A change in one base within a triplet of bases on DNA is called a **single gene mutation**.

Types of single gene mutation

Substitution – One base (nucleotide) within a triplet of DNA bases is substituted for another base, coding for one different amino acid in the polypeptide chain. This causes a minor change to the finished protein.

Insertion – An additional base is inserted into a triplet of bases altering the base sequence of the gene to the right of the insertion. Many base triplets coding for many amino acids are affected, causing a major change in the finished protein. This is called a **frameshift mutation**.

Deletion – A base is deleted from a triplet of bases altering the base sequence to the right of the deletion, as all triplets move along one space to fill the gap. Many base triplets coding for many amino acids are affected, causing a major change to the structure of the finished protein. This is called a **frameshift mutation**.

Part of original DNA chain

DNA base sequence	CCG	GTG	TCA	TGT	GGT	AAG
mRNA codons	GGC	CAC	AGU	ACA	CCA	UUC
resulting amino acid sequence	glycine	histidine	serine	threonine	proline	phenylalanine

Effect of substitution

A replaces T here		↓				
DNA base sequence	CCG	GAG	TCA	TGT	GGT	AAG
mRNA codons	GGC	CUC	AGU	ACA	CCA	UUC
resulting amino acid sequence	glycine	leucine	serine	threonine	proline	phenylalanine

Effect of inversion

G and T reversed here		↓				
DNA base sequence	CCG	GGT	TCA	TGT	GGT	AAG
mRNA codons	GGC	CCA	AGU	ACA	CCA	UUC
resulting amino acid sequence	glycine	proline	serine	threonine	proline	phenylalanine

Effect of deletion

G lost from original chain here		↓	all bases moved one place to the left			
DNA base sequence	CCG	TGT	CAT	GTG	GTA	AG
mRNA codons	GGC	ACA	GUA	CAC	CAU	UC
resulting amino acid sequence	glycine	cysteine	valine	histidine	histidine	

Effect of insertion

additional G added here	↓			all bases moved one place to the left			
DNA base sequence	CCG	GGT	GTC	ATG	TGG	TAA	G
mRNA codons	GGC	CAC	AGU	ACA	CCA	UUC	
resulting amino acid sequence	glycine	proline	glutamine	tyrosine	threonine	isoleucine	

Nucleotide sequence repeat

A **repeat expansion** mutation can increase the number of times a short DNA sequence of three or four base pairs is repeated. This can affect the function of the protein that is coded for by a gene.

TOP TIP

Remember types of point mutation as 'SID':
• **S**ubstitution
• **In**sertion
• **D**eletion.

Single base (nucleotide) substitutions

• **Missense** – One triplet (codon) of bases is changed, now coding for an incorrect amino acid.

• **Nonsense** – One triplet is changed into a 'stop' codon, stopping the formation of the polypeptide chain.

• **Splice site mutation** – A codon located at a site of intron splicing on mRNA is changed, allowing some introns to remain within the primary mRNA transcript.

Quick Test 10

1. Name the two parts of a genome.
2. What is the function of non-protein coding sequences?
3. Name two mutagenic agents which increase the frequency of mutation.
4. What is meant by the term 'single gene mutation'?
5. Explain the term 'frameshift mutation'.
6. In what way does a 'splice site mutation' affect the mature mRNA transcript?

Chromosome structure mutations

Change in chromosome number

A mutation can occur during the formation of human egg cells in the ovary, which causes some egg cells to have one too many chromosomes and some to have one too few. This type of mutation causes a change in **chromosome number**. Normal egg cells contain 23 chromosomes.

If an egg with 24 chromosomes is fertilised by a normal sperm cell containing 23 chromosomes, the resulting embryo will have 47 chromosomes within the body cells instead of 46. This is a genetic condition called **Down's Syndrome**.

Human eggs that have one chromosome missing are not viable and are unlikely to become fertilised.

Change in chromosome structure

If whole sections of chromosomes containing many genes are broken, rearranged or lost, this is a mutation of **chromosome structure**.

- **Deletion** – This occurs when two breaks occur along the length of the chromosome, and the middle segment of chromosome containing many genes is lost. The broken or 'sticky ends' of the remaining chromosome join together.

- **Duplication** – This occurs when a broken segment from a similar neighbouring chromosome is inserted, duplicating a specific set of genes.

- **Inversion** – This is when two breaks occur along the length of a chromosome. The segment of chromosome between the breaks rotates 180 degrees and reattaches, reversing the gene sequence.

- **Translocation** – This occurs when a segment breaks off the end of one chromosome and attaches to the end of a neighbouring chromosome, adding additional genes.

> **TOP TIP**
>
> Remember 'DDTI' for types of chromosome mutation:
> - **D**eletion
> - **D**uplication
> - **T**ranslocation
> - **I**nversion.

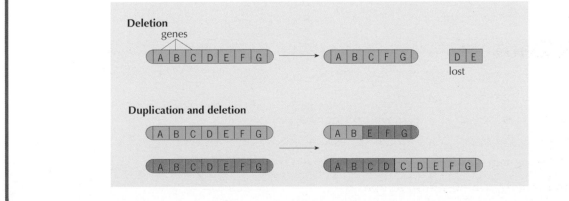

Inversion

```
( A  B  C  D  E  F  G )  ⟶  ( A  B  C  F  E  D  G )
            └────┴───┘
         this section is inverted
```

Translocation

```
( A  B  C  D  E  F  G )              ( A  B  P  Q  R  S )

                          ⟶

( L  M  N  O  P  Q  R  S )           ( L  M  N  O  C  D  E  F  G )
```

Polyploidy

Some cells can receive **double the normal number of chromosomes** usually present in the nucleus due to a fault that occurs during cell division. This results in cells containing **whole genome duplications**, and they are referred to as being **polyploid**.

Polyploid plants are larger than normal plants and produce much larger fruits or seeds.

Normal fruit

Polypoid fruit

Polyploid animals

Polyploidy occurs in some animals as a result of a fault during the production of gametes. Polyploidy in animals is rare compared to polyploidy in plants.

Yellow spotted salamander

Quick Test 11

1.	Give an example of a genetic condition in humans caused by a change in chromosome number.
2.	Name the four types of chromosome structure mutation.
3.	Which type of chromosome structure mutation results in a segment of one chromosome attaching to the end of a neighbouring chromosome?
4.	What does the term 'polyploidy' mean?
5.	What advantage do polyploid plants have compared to normal plants?

Evolution

Evolution is the gradual change in inherited characteristics (variation within the genome) within populations of living organisms, over hundreds of generations.

Inheritance

Types of gene transfer (inheritance of genetic characteristics):

1. **Vertical** – The genetic code on DNA is transferred from parent to offspring, as a result of asexual or sexual reproduction.

2. **Horizontal – prokaryote to prokaryote** – In prokaryotes, genetic information (DNA) is passed from cell to cell, resulting in fast evolutionary change.

3. **Horizontal – prokaryote to eukaryote** – Bacteria such as agrobacterium and viruses can transfer their DNA directly into eukaryote cells, where it is incorporated into the DNA of the eukaryote cell. Horizontal gene transfer between agrobacterium and plant cells is useful in genetic engineering.

Natural selection

Natural selection is a **non-random** increase in the frequency of those specific genes which increase an organism's chance of survival.

Darwin's Theory of Natural Selection:

1. Organisms produce more offspring than the environment can support.

2. Genetic variation occurs within individuals of a population.

3. Individuals compete for available resources such as food and mates.

4. Individuals with favourable genes which give them an **advantage** in the environment, such as the ability to find prey, are more likely to survive and pass genes on to the next generation.

5. Frequency of these favourable genes increases within the population, individuals **without** these favourable genes gradually begin to die out.

Sexual selection

The chances of successful reproduction are increased if a sexual partner is selected based on the recognition and **non-random** selection of strong genetic traits.

1. Male/male rivalry – Males compete with physical aggression for access to females. For example, male deer fighting with antlers.

2. Female choice – Selection of a male based upon a high quality visual display of genetic traits. For example, the eye spots on the feathers of male peacocks. The larger the eyespots, the more attractive the male peacock is to females.

Mechanisms of natural selection for a measurable genetic trait

Natural selection can affect the frequency of a measureable trait, such as height or mass, within a large population in three ways.

1. **Stabilising selection** occurs within a stable environment, and selects for the more numerous 'average' versions of a genetic trait within a population, avoiding the more extreme versions of the trait.

2. **Directional selection** occurs within a changing environment, and selects for a less common version of a genetic trait within a population, resulting in an increase in frequency.

3. **Disruptive selection** occurs when two different types of environment or resources become available, and selects for the most extreme versions of a genetic trait, resulting in the population splitting into two distinct groups.

Genetic drift

A **random** increase or decrease in the frequency of DNA sequences within the **gene pool** of a population, as a result of neutral mutations and founder effects, is called **genetic drift**.

Founder effect

Specific alleles of genes may be lost from a population when some organisms become isolated from the main population and 'found' a new separate population. The new population contains a **random selection** of alleles from the parent population. This is called the **founder effect**.

Bottleneck effect

The majority of a population may die out due to a natural disaster such as tsunami, flood or fire and a large number of genes are lost. However, survivors carry a small random selection of genes from the original population. This results in a change in the frequency of alleles in a population and significant loss of genetic variation.

Original population

New population

50:50 allele frequency

Population passes through a bottleneck: only a few survive
By chance, there is 70:30 allele frequency

70:30 allele frequency

Speciation

The evolution of a new biological species as a result of isolation, mutation and natural selection is called **speciation**.

1. **Allopatric speciation** – A population is split into two sub-populations, A and B, which are separated by a geographical barrier, preventing gene flow. Mutations occur in populations A and B, followed by natural selection over many generations, resulting in two new species A and B which can now no longer interbreed.

2. **Sympatric speciation** – Two sub-populations become isolated from the original population due to behavioural or ecological barriers, or by polyploidy occurring within a population. This prevents gene flow between the two sub-populations which begin to evolve along different routes, and can eventually become two distinct species.

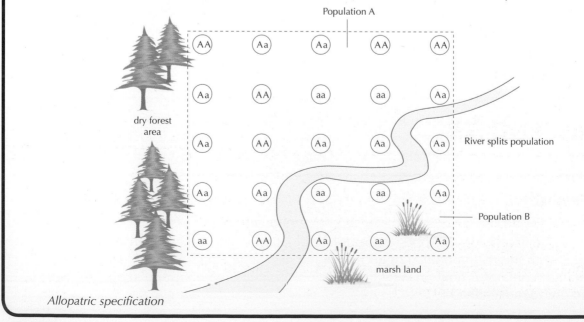

Population A

dry forest area

River splits population

Population B

marsh land

Allopatric specification

Hybrid zone

Several populations of an organism may occupy the same area and vary in their ability to interbreed. The area where two populations overlap and breeding takes place between them is known as a **hybrid zone**. All interbreeding populations are connected through interbreeding regions or hybrid zones.

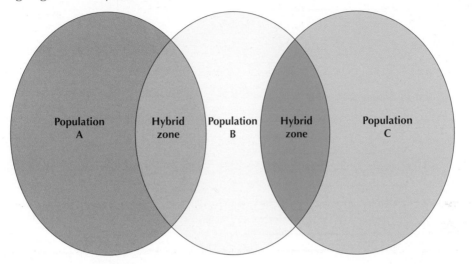

If a hybrid zone is narrow, this could indicate that the two overlapping populations are genetically distinct species. A wide hybrid zone may indicate greater genetic variation within the overlapping populations.

TOP TIP

To help remember the types of genetic selection involved in the process of evolution, try drawing a mind map diagram. It summarises the information in a visual way, and allows you to see and remember important connections.

Quick Test 12

1. Give one advantage of horizontal gene transfer in prokaryotes.
2. Why might some individuals within a population fail to survive and reproduce?
3. What is the advantage of sexual selection to a population?
4. What is meant by the term genetic drift?
5. Which types of barriers are involved in sympatric speciation?
6. Describe the 'founder effect'.

Genomic sequencing

The analysis of the sequence of bases within an organism's DNA is called **genomic sequencing**, and is carried out within a laboratory. This process involves **bioinformatics**, computer technology that stores, organises and statistically analyses sequencing information.

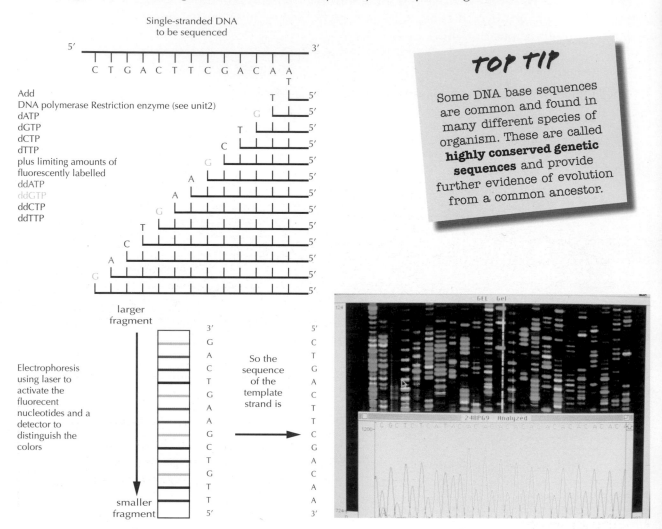

TOP TIP

Some DNA base sequences are common and found in many different species of organism. These are called **highly conserved genetic sequences** and provide further evidence of evolution from a common ancestor.

Genome sequencing can be used to study evolutionary relationships between different species by examining similarities and differences between base sequences on their DNA in an area of research called **phylogenetics**. Species that are closely related have similar genomic sequences. Some genetic sequences appear to be common across many different species.

Molecular clocks

As DNA and RNA molecules age, the frequency of base substitution mutations increases. The point in time when two species of organisms, which both evolved from a common ancestor, began to diverge can be calculated. This is done by comparing the base substitutions within each species, which is proportional to the length of time elapsed since the two species began to diverge away from each other.

Information from both phylogenetics and molecular clocks can be used to plot the sequence of events that occurred during the evolution of life on Earth, from prokaryotes to multicellular organisms. This information has lead to the classification of life on Earth into three main groups.

Three domains of living organisms

These have been determined by genomic sequencing techniques:

1. Bacteria (prokaryotes)
2. Archaea (prokaryotes living in extreme conditions of heat or salinity)
3. Eukaryotes – plants, animals and fungi.

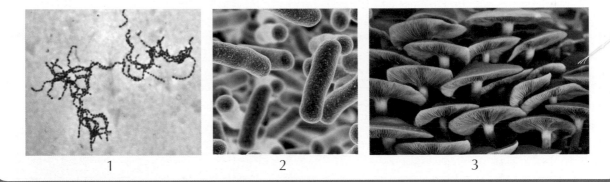

1   2   3

Personal genomics

Following the completion of the Human Genome Project in 2003, the DNA of a person can now be sequenced providing personal information about genetic disorders, susceptibility to particular diseases and the possible reaction of the body to certain drugs. This means that it is now possible to personalise medical treatment.

Ethical issues

- Informing a person of possible future illnesses may cause distress.
- Employers and insurance companies may avoid people who they suspect may become ill in the future if they have access to this information.

Quick Test 13

1.	What kind of information does genomic sequencing provide?
2.	How does bioinformatics contribute to this process?
3.	List the three domains of living organisms.
4.	In what way is a molecular clock useful in the study of the evolution of two different species from a common ancestor?
5.	State the advantages and disadvantages of personal genomics.

Metabolic pathways

Chains of different biochemical reactions called **metabolic pathways** make up the **metabolism** of a cell.

Anabolic and catabolic

Anabolic pathways make or **synthesise** new molecules from basic building blocks using energy in the form of adenosine triphosphate (ATP), for example, building up protein molecules from amino acids.

Catabolic pathways break down large complex molecules into their smaller subunits, with the release of energy (ATP). For example, the process of cell respiration involves the breakdown of glucose with oxygen to form ATP, carbon dioxide and water.

> **TOP TIP**
>
> Anabolic reactions – energy IN
> Catabolic reactions – energy OUT
> (Remember: put **out** the **cat**!)

Integrated metabolic pathways

Anabolic and catabolic pathways can become interdependent and form an integrated metabolic pathway. The catabolic pathway of cell respiration producing ATP is linked to the anabolic pathway of synthesis of proteins from amino acids, which requires energy.

Reversible and irreversible steps

Each step of a metabolic pathway is controlled by an enzyme. Some steps are reversible, meaning that metabolite A may be converted to metabolite B, and metabolite B can be converted back again to metabolite A and used perhaps in an alternative pathway. The cell has control over the pathway.

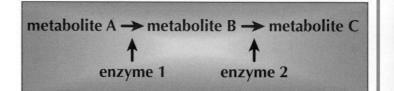

An irreversible step results in one metabolite being fully converted to another, with no alternative pathway available.

Alternative routes

Specific steps in a pathway can be bypassed using an alternative route or 'short cut', so that metabolite A is converted directly to metabolite C, for example.

Cell membrane

The cell membrane controls the entry and exit of molecules in a cell. It is also the site of many metabolic reactions. The cell membrane is <u>semi-permeable</u>, allowing only <u>some</u> molecules to pass through.

Any structure within the cell cytoplasm is called a **cell organelle**. For example, the nucleus, mitochondrion, chloroplast and vacuoles are all cell organelles.

Mitochondria have a folded inner membrane providing compartments which isolate or bring together specific metabolites within a pathway:

- inner membrane folds – site of ATP production
- space between folds (central matrix) – contains metabolites and enzymes for the citric acid cycle in cell respiration.

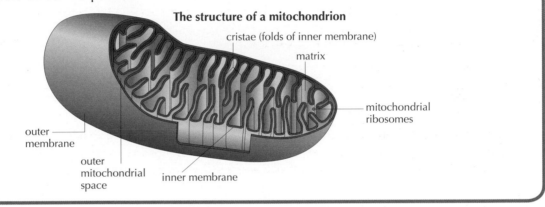

The structure of a mitochondrion

cristae (folds of inner membrane)

matrix

mitochondrial ribosomes

outer membrane

outer mitochondrial space

inner membrane

Structure of the cell membrane

The cell membrane or **plasma membrane** is formed from a dynamic bilayer of phospholipid molecules, interspersed with different types of protein molecule. There are no chemical bonds between the phospholipids and protein molecules, or between the phospholipids themselves, only weak electrostatic forces. The plasma membrane also contains cholesterol which helps maintain fluidity and polysaccharide molecules, which help cells to attach to each other forming a tissue.

This type of membrane structure diagram is called the **fluid mosaic model**.

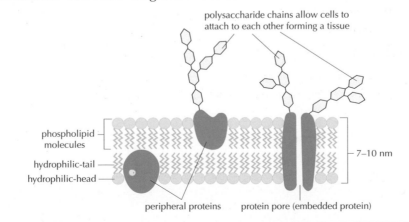

polysaccharide chains allow cells to attach to each other forming a tissue

phospholipid molecules

hydrophilic-tail

hydrophilic-head

7–10 nm

peripheral proteins

protein pore (embedded protein)

Methods of molecular transport

1. **Diffusion** is a movement of molecules through the phospholipids bilayer **from a high** concentration of molecules **to a low** concentration. Diffusion is a **passive process**, it does not require energy (ATP).

2. **Osmosis** is a movement of water molecules only **from a region of high water concentration** to a **low water concentration** through the semi-permeable plasma membrane. Osmosis is also a passive process.

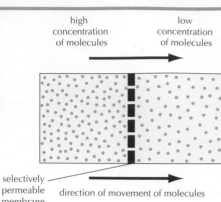

high concentration of molecules

low concentration of molecules

selectively permeable cell membrane

direction of movement of molecules

Before osmosis

selectively permeable

solution A

solution B

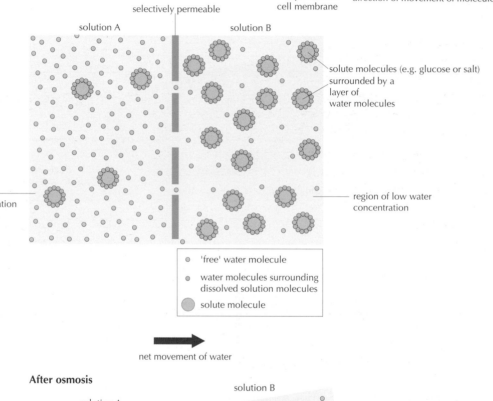

solute molecules (e.g. glucose or salt) surrounded by a layer of water molecules

region of high water concentration

region of low water concentration

- ○ 'free' water molecule
- ● water molecules surrounding dissolved solution molecules
- ⬤ solute molecule

net movement of water

After osmosis

solution B

solution A

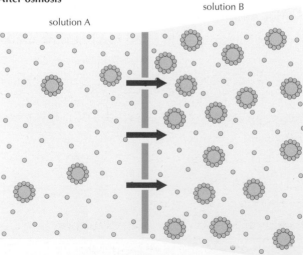

Solution A is now <u>isotonic</u> to solution B. They have the same concentration of water molecules.

3. **Protein pores** are 'channels' in large protein molecules which span the plasma membrane. They act as 'gateways' for large molecules to pass through by diffusion.

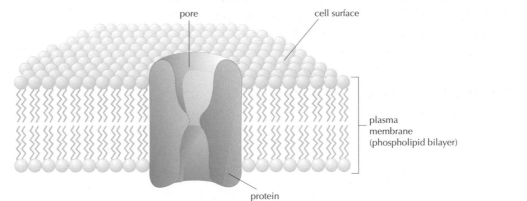

pore   cell surface

plasma membrane (phospholipid bilayer)

protein

4. **Protein pump** is a carrier protein on the plasma membrane which moves (pumps) molecules and ions across the membrane from a **low** concentration to a **high** concentration (the opposite of diffusion). The movement of the carrier protein **requires energy** in the form of ATP. This is an <u>active</u> process.

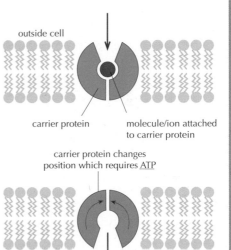

outside cell

carrier protein   molecule/ion attached to carrier protein

carrier protein changes position which requires <u>ATP</u>

5. **Sodium potassium pump** is a carrier protein on the plasma membrane, which transports sodium ions **out** of the cell and transport potassium ions **in**. This **requires energy** in the form of ATP. This is an <u>active</u> process.

inside cell

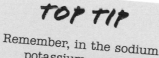

TOP TIP

Remember, in the sodium potassium pumps:
• **s**odium ions are **s**ent out.
• **P**otassium ions are **p**ulled in!

Quick Test 14

1.	Name two different types of metabolic pathway.
2.	What is the advantage to a cell of having a reversible step within a metabolic pathway?
3.	Why is the compartmentalisation of the inner membrane of a mitochondrion important in terms of metabolic pathways?
4.	Which two molecules make up the structure of the cell membrane?
5.	What is the function of the cell membrane?
6.	Describe what is meant by the term 'cell organelle'.
7.	How does the process of active transport differ from diffusion?
8.	What is the function of protein pores in the plasma membrane?
9.	Which two methods of molecular transport involve a carrier protein?

Enzyme control of metabolic pathways

Each step within a metabolic pathway is controlled by one specific enzyme, coded for by one gene. If an enzyme within the pathway is missing, or does not work properly due to mutation, the chain of chemical reactions is disrupted.

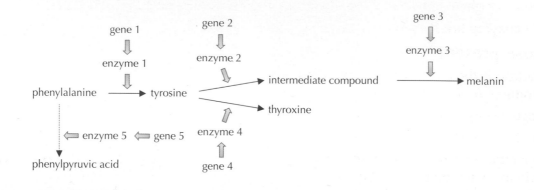

Jacob Monod hypothesis

Researchers Jacob and Monod investigated the mechanism by which a strain of E.Coli bacteria is able to control a specific metabolic pathway by switching a gene 'on' and 'off'.

- Gene 'switched on' – actively coding for enzyme – enzyme produced.
- Gene 'switched off' – not actively coding for enzyme – no enzyme produced.

E.Coli can break down **lactose** sugar into glucose and galactose by producing the enzyme beta galactosidase. The bacterial cells then use the glucose for cell respiration. The enzyme is only produced when **lactose is present**. This prevents wastage of valuable cell resources such as ATP and amino acid molecules.

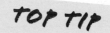

TOP TIP

The Jacob Monod hypothesis is an excellent example to use if answering a question about genetic control of a metabolic pathway.

Enzyme induction

The gene coding for the enzyme **B-galactosidase** is 'switched on' only when **lactose is present**. This is called **enzyme induction**. The mechanism of enzyme induction involves three genes on the bacterial DNA.

1. **Structural gene** – codes for enzyme.
2. **Operator gene** – switches structural gene 'on' and 'off'.
3. Regulator gene – codes for **'repressor'** protein molecules.

The 'structural' and 'operator' genes together are called the **operon**.

Mechanism of genetic control in E.Coli

Lactose absent

- Regulator gene codes for protein repressor molecule.
- Repressor molecule attaches to operator gene.
- Operator gene **deactivated** – cannot switch on structural gene.
- No enzyme produced.

Lactose present

- Lactose is the **inducer** molecule, stimulating the production of the enzyme.
- Regulator gene codes for repressor molecules.
- Repressor molecules combine with lactose.
- Operator gene now free of repressor molecules and is able to switch on structural gene.
- Structural gene codes for and produces B-galactosidase.
- B-galactosidase breaks down lactose/repressor complex into glucose and galactose, **releasing repressor molecules**.
- Free repressor molecule attaches to operator gene, switching off structural gene.
- No enzyme produced, all lactose has been digested.

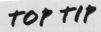

TOP TIP

E.Coli cannot use lactose to produce energy in cell respiration. It must be broken down into glucose and galactose. Glucose can be used by the cell for cell respiration to produce ATP.

TOP TIP

Lactose is the sugar found in fresh milk. Some babies are 'lactose intolerant' which means they have difficulty digesting milk and are given formula milk without lactose instead.

If lactose is absent –

1. The regulator gene codes for the production of repressor protein molecules.

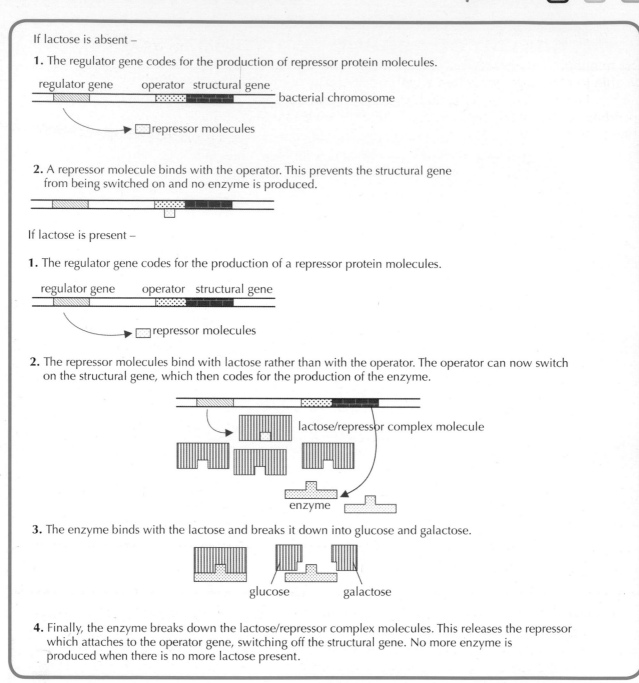

2. A repressor molecule binds with the operator. This prevents the structural gene from being switched on and no enzyme is produced.

If lactose is present –

1. The regulator gene codes for the production of a repressor protein molecules.

2. The repressor molecules bind with lactose rather than with the operator. The operator can now switch on the structural gene, which then codes for the production of the enzyme.

3. The enzyme binds with the lactose and breaks it down into glucose and galactose.

4. Finally, the enzyme breaks down the lactose/repressor complex molecules. This releases the repressor which attaches to the operator gene, switching off the structural gene. No more enzyme is produced when there is no more lactose present.

Quick Test 15

1. Describe the effect of a mutation in a gene which controls a specific step within a metabolic pathway.

2. What is meant by the term 'enzyme induction' within the Jacob Monod hypothesis?

3. Name the three genes involved in the mechanism of enzyme induction within E.Coli.

4. Which gene codes for the production of the enzyme B-galactosidase?

5. Which gene codes for repressor molecules?

6. Identify the 'inducer molecule' in the Jacob Monod example of genetic control.

Enzyme action

The minimum energy required for chemical molecules to be able to react together within a cell is called **activation energy**. Enzymes are catalysts that **lower** activation energy, speeding up chemical reactions that would otherwise be too slow to sustain life.

TOP TIP

Enzymes speed up chemical reactions within the cell, but do not take any part in the reaction, remaining unchanged.

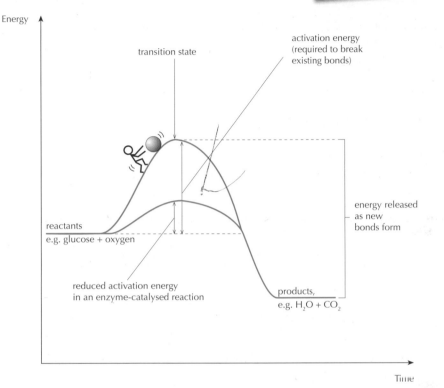

Energy

transition state

activation energy (required to break existing bonds)

energy released as new bonds form

reactants
e.g. glucose + oxygen

reduced activation energy in an enzyme-catalysed reaction

products, e.g. $H_2O + CO_2$

Time

Enzyme structure

Enzymes are three dimensional globular protein molecules that are sensitive to both pH and temperature. The specific part of the enzyme molecule that connects with a substrate molecule is called the **active site**.

The **shape** of the active site on an enzyme molecule is specific to only **one** substrate. If the shape of a substrate molecule is complementary to the active site of the enzyme, it will have a **high affinity** for the active site. Products of enzyme substrate reactions have a **low affinity** for the active site of the enzyme.

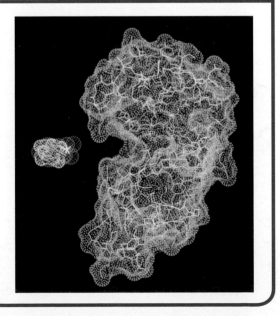

TOP TIP

The active site of an enzyme is complementary to the shape of a specific substrate molecule. Imagine two jigsaw pieces of the correct complementary shapes being able to fit together.

Induced fit

Substrate molecules that have a similar shape (but not exact complementary) to the active site of a specific enzyme may still have some affinity and be able to react. The active site can alter shape slightly to accommodate an alternative substrate molecule. This is called **induced fit**.

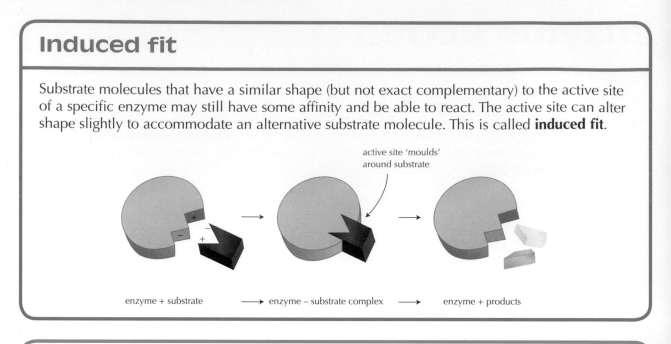

active site 'moulds' around substrate

enzyme + substrate   ⟶   enzyme – substrate complex   ⟶   enzyme + products

Control of rate of enzyme reactions

Substrate concentration – Increasing substrate concentration increases the rate of reaction until all active sites on the enzyme molecule are full, when the rate of reaction remains constant.

End product concentration – As the concentration of the end product of an enzyme substrate reaction within a metabolic pathway increases, some end product molecules bind to the enzyme 1 catalysing the first step in the pathway. This slows down the conversion of metabolite A to B in the pathway, meaning less end product is produced releasing enzyme 1, increasing the conversion of A to B producing more end product and so on. This is an example of **negative feedback control** of a metabolic pathway.

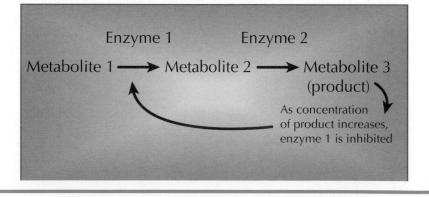

Enzyme 1   Enzyme 2

Metabolite 1 ⟶ Metabolite 2 ⟶ Metabolite 3 (product)

As concentration of product increases, enzyme 1 is inhibited

Multi-enzyme complexes

Groups of enzymes working together have more than one function. They are regulated by multiple substrates and **allosteric effectors**, and are able to catalyse faster reactions than if they were operating within separate metabolic pathways.

Orientation of active site and substrate molecules

The shape of the active site determines the position (orientation) of the two substrate molecules during a synthesis reaction. This reduces the activation energy. The product molecules have a low affinity for the active site, and are unlikely to compete with the substrate molecules.

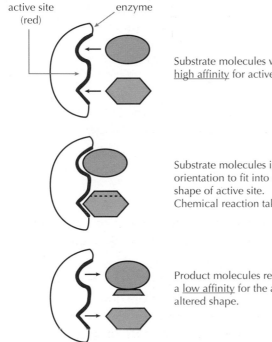

active site (red)   enzyme

Substrate molecules with <u>high affinity</u> for active site.

Substrate molecules in correct orientation to fit into complimentary shape of active site.
Chemical reaction takes place

Product molecules released, each with a <u>low affinity</u> for the active site, due to altered shape.

TOP TIP

Synthesis reaction – Enzyme joins two molecules together to make the product.
Degradation reaction – Enzyme breaks down a large substrate molecule into two smaller product molecules.

Quick Test 16

1. What effect does lowering the activation energy have on the rate of a chemical reaction?
2. Describe the principle of induced fit in an enzyme substrate reaction.
3. Describe the effect of increasing substrate concentration on the rate of reaction.
4. What is the advantage of a multi-enzyme complex to the cell?
5. What effect does the orientation of substrate molecules in the active site of an enzyme have on the resulting enzyme substrate reaction?

Enzyme inhibition

The reaction between enzyme and substrate molecules can be affected by the action of an **inhibitor**. Inhibitor molecules can slow down metabolic pathways which are controlled by enzymes.

There are three types of inhibitor molecule:

1. Competitive inhibitor
2. Non-competitive inhibitor
3. End product inhibitor (feedback inhibition).

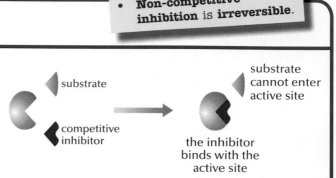

TOP TIP

Remember:
- **Competitive** inhibition is **reversible**
- **Non-competitive** inhibition is **irreversible**.

Competitive inhibitor

Competitive inhibitor molecules slow down the rate of reaction by actively competing with substrate molecules for the active site of an enzyme. If the active site contains an inhibitor molecule, the substrate molecule cannot enter.

substrate

competitive inhibitor

substrate cannot enter active site

the inhibitor binds with the active site

The effect of the inhibitor can be reduced by increasing the concentration (number) of substrate molecules, which increases the reaction rate. Substrate molecules outnumber inhibitor molecules and are more likely to win the race for the active sites!

Increasing the concentration of enzyme (number of active sites) can also reduce the effect of a competitive inhibitor. **Competitive inhibition is therefore reversible**.

Non-competitive inhibitor

Non-competitive inhibitor molecules do not attach to the active site, but to another part of the enzyme molecule called the **allosteric site**. Once in position, the inhibitor changes the three dimensional structure of the active site preventing connection with substrate molecules, and decreasing the rate of reaction. Altering the enzyme or substrate concentration has no effect. **Non-competitive inhibition is irreversible**.

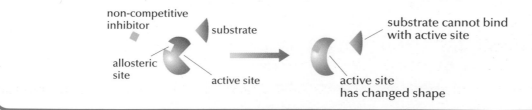

non-competitive inhibitor

substrate

allosteric site

active site

substrate cannot bind with active site

active site has changed shape

End product inhibitor

End product inhibition can control a metabolic pathway, where the end product acts as a non-competitive inhibitor. This happens because when levels of end product molecules become too high, inhibition of the enzymes in the pathway occurs, stopping production. When end product levels fall, inhibition is removed and production begins again. This is called **feedback inhibition**.

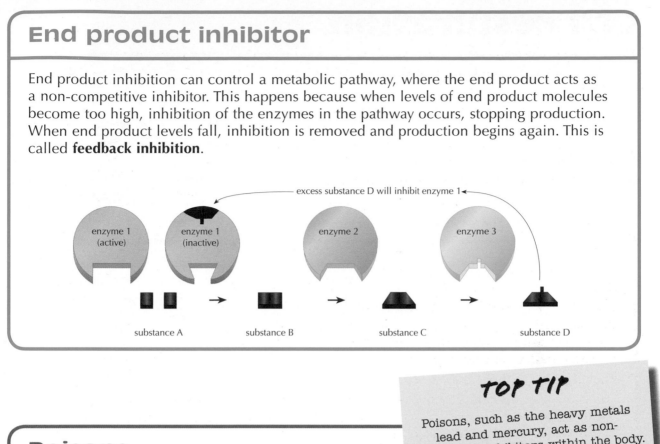

Poisons

TOP TIP

Poisons, such as the heavy metals lead and mercury, act as non-competitive inhibitors within the body.

Poisons such as arsenic, and heavy metals such as lead and mercury are non-competitive inhibitors. They act by attaching to the allosteric site of an enzyme, and destabilising the 3D structure of the molecule. The active site changes shape, and is unable to connect with the substrate molecule.

Quick Test 17

1. What is the effect of an inhibitor molecule on the rate of reaction between an enzyme and substrate?
2. Which type of inhibitor alters the shape of the active site of an enzyme?
3. How can the effects of a competitive inhibitor be reduced?
4. Give two examples of substances which act as non-competitive inhibitors if they enter the body.
5. What occurs at the allosteric site of an enzyme?
6. In what way does end product inhibition benefit a cell?

Cell respiration

The process of life

The process of life at the most basic level, is a flow of electrons.

'You eat sugars that have excess electrons, and you breathe in oxygen that willingly takes them.' K. Nealson, *New Scientist*, July 2014.

Adenosine triphosphate (ATP)

ATP is a high energy chemical compound needed by cells to do 'work', such as DNA replication, cell division, active transport, synthesis of new molecules and muscle contraction.

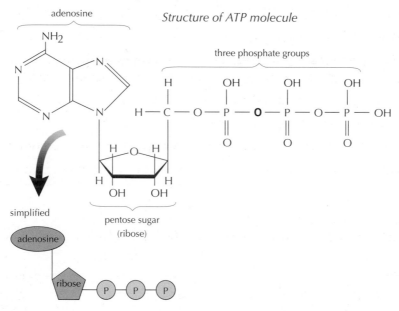

Structure of ATP molecule

When the terminal phosphate of an ATP molecule is broken off by the enzyme ATPase, the stored up (potential) energy in the chemical bond is released for the cell to use as fuel. What is left is the low energy compound adenosine diphosphate (ADP) and an inorganic phosphate molecule.

Using energy from the breakdown of glucose in a cell, an inorganic phosphate molecule can be reattached onto ADP to regenerate ATP. This is called **phosphorylation**.

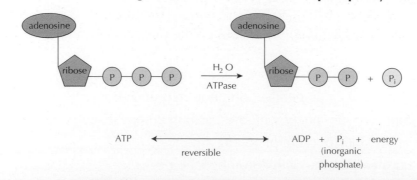

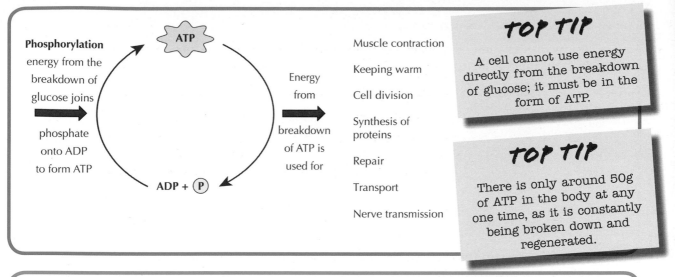

TOP TIP

A cell cannot use energy directly from the breakdown of glucose; it must be in the form of ATP.

TOP TIP

There is only around 50g of ATP in the body at any one time, as it is constantly being broken down and regenerated.

Metabolic pathways in cell respiration

Cell respiration is the breakdown (oxidation by the removal of hydrogen atoms) of glucose molecules within a cell to release energy for the phosphorylation of ADP + Pi to ATP.

There are three pathways in the process of cell respiration.

1. **Glycolysis** which occurs in the cytoplasm.
2. **Citric acid cycle** which occurs in the central matrix of the mitochondria.
3. **Electron transport chain** which occurs on the cristae of the mitochondria.

Glycolysis

A 6 carbon glucose molecule is broken into two molecules of a 3 carbon compound called **pyruvate**. Two molecules of ATP are needed to break the glucose molecule. This is called the **energy investment phase**.

When the chemical bonds within the glucose molecule are broken, enough stored up (potential) energy is released to produce 4 ATP molecules. This is called the **energy payoff phase**.

During glycolysis, the cell has a **net energy gain of 2 ATP molecules**.

Hydrogen ions are removed from the broken glucose molecule by the enzyme **dehydrogenase** during the 'energy payoff' stage. As hydrogen ions are extremely unstable within a living cell, they are immediately picked up by a hydrogen carrier (co-enzyme) molecule called NAD to form NADH.

TOP TIP

Glycolysis:
'glyco' – means 'glucose',
'lysis' – means 'to break'.

TOP TIP

The chemical formula for glucose is $C_6H_{12}O_6$. It is called a 'carbohydrate' because it contains the elements carbon, hydrogen and oxygen.

Fermentation

If a cell has **no oxygen**, pyruvate is converted to **lactic acid** in **animal cells** and **ethanol** (alcohol) in **plant cells**, together with the release of carbon dioxide.

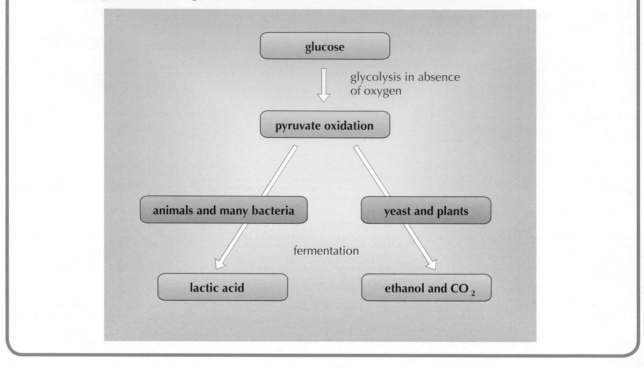

Quick Test 18

1. Give three examples of functions carried out by a cell that require energy in the form of ATP.

2. Write a word equation to show the process of phosphorylation.

3. Explain why there is only a very small mass of ATP present in the body at any one time.

4. Name the hydrogen carrier in cell respiration.

5. What is the name of the 3 carbon compound produced in glycolysis?

6. Where does glycolysis take place in a cell?

7. Explain why a cell receives a net gain of 2 ATP during glycolysis.

Citric acid cycle and electron transport chain

Glycolysis takes place in the cytoplasm of a cell. The citric acid cycle and electron transport chain take place in the mitochondria. The citric acid cycle occurs in the matrix of the mitochondria, and the electron transport chain occurs in the folds of the inner membrane called the cristae.

outer 'plasma' membrane

'central' matrix

inner folds of membrane

cristae

10 μm

Citric acid cycle

The citric acid cycle takes place in the central matrix of the mitochondrion, and occurs **only if a cell has oxygen**.

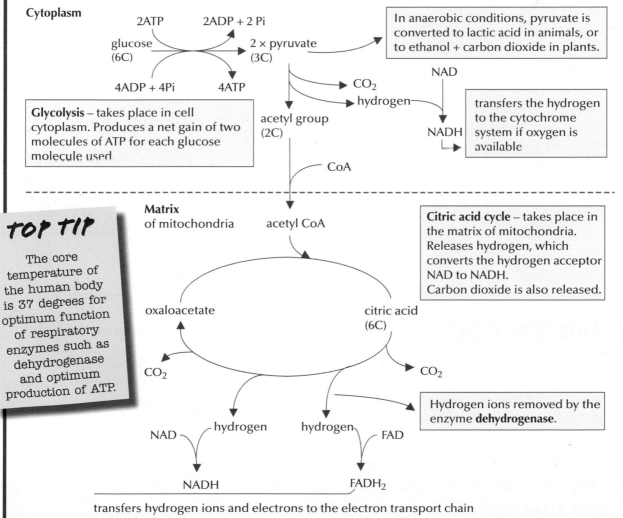

Cytoplasm

$2ATP$ $2ADP + 2\,Pi$

glucose (6C)   2 × pyruvate (3C)

$4ADP + 4Pi$ $4ATP$

In anaerobic conditions, pyruvate is converted to lactic acid in animals, or to ethanol + carbon dioxide in plants.

CO_2

hydrogen

NAD

Glycolysis – takes place in cell cytoplasm. Produces a net gain of two molecules of ATP for each glucose molecule used.

acetyl group (2C)

NADH — transfers the hydrogen to the cytochrome system if oxygen is available

CoA

Matrix of mitochondria

acetyl CoA

Citric acid cycle – takes place in the matrix of mitochondria. Releases hydrogen, which converts the hydrogen acceptor NAD to NADH. Carbon dioxide is also released.

oxaloacetate

citric acid (6C)

CO_2 CO_2

Hydrogen ions removed by the enzyme **dehydrogenase**.

NAD — hydrogen   hydrogen — FAD

NADH $FADH_2$

transfers hydrogen ions and electrons to the electron transport chain

TOP TIP

The core temperature of the human body is 37 degrees for optimum function of respiratory enzymes such as dehydrogenase and optimum production of ATP.

- Pyruvate (from glycolysis) is converted to an acetyl group which combines with co-enzyme A to form **acetyl co-enzyme A (acetyl-CoA)** which enters the citric acid cycle.
- Dehydrogenase removes hydrogen ions which combine with NAD to form NADH.
- Acetyl co-enzyme A (2 carbon compound) combines with **oxaloacetate** (4 carbon compound) to form **citrate** (6 carbon compound).
- Several enzyme controlled steps occur within the cycle resulting in the regeneration of more oxaloacetate.
- Carbon dioxide is released, ATP molecules generated and NADH formed.
- A second co-enzyme FAD (flavine adenine dinucleotide) accepts hydrogen ions and electrons to become $FADH_2$.

Electron transport chain

The electron transport chain takes place on the cristae (inner folds) of the mitochondria. Here a chain of protein carrier molecules accept high energy electrons from the hydrogen carriers NADH and $FADH_2$.

- High energy electrons flow along the chain of protein electron carrier molecules.
- Energy released from the flow of electrons is used to pump hydrogen ions along the inner membrane of the mitochondria.
- The flow of hydrogen ions **rotates** part of a membrane protein called **ATP synthase**, phosphorylating ADP + Pi to ATP.
- When the low energy hydrogen ions and electrons reach the end of the protein carrier chain they combine with oxygen, which is called the **final hydrogen acceptor**, to form water.

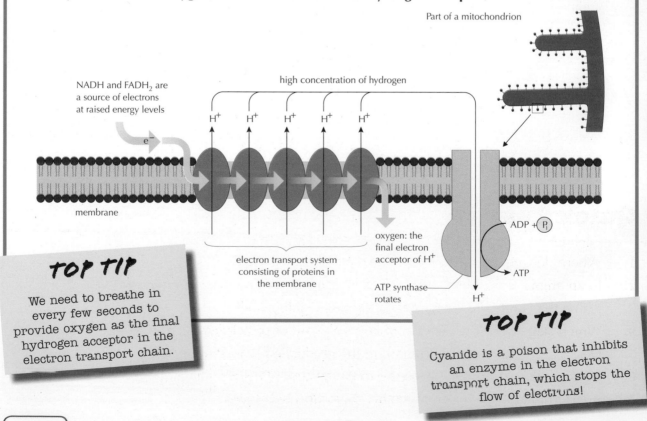

Part of a mitochondrion

NADH and $FADH_2$ are a source of electrons at raised energy levels

high concentration of hydrogen

H^+ H^+ H^+ H^+ H^+

e^-

membrane

electron transport system consisting of proteins in the membrane

oxygen: the final electron acceptor of H^+

ATP synthase rotates

H^+

ADP + Pi

ATP

TOP TIP

We need to breathe in every few seconds to provide oxygen as the final hydrogen acceptor in the electron transport chain.

TOP TIP

Cyanide is a poison that inhibits an enzyme in the electron transport chain, which stops the flow of electrons!

Alternative respiratory substrates

Starch in plant cells and glycogen in animal cells are storage carbohydrates, which can be broken down by enzymes into glucose for cell respiration.

Fats can be broken down into fatty acids and glycerol, both of which may be metabolised through glycolysis and the citric acid cycle.

Proteins can be broken down by the digestive enzyme pepsin into amino acids, some of which can be broken down further into urea and intermediate compounds which in turn form pyruvate, and enter the citric acid cycle.

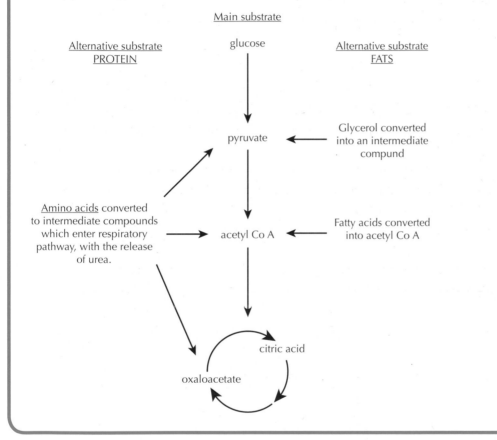

Main substrate

glucose

Alternative substrate
PROTEIN

Alternative substrate
FATS

pyruvate ← Glycerol converted into an intermediate compund

Amino acids converted to intermediate compounds which enter respiratory pathway, with the release of urea.

acetyl Co A ← Fatty acids converted into acetyl Co A

citric acid

oxaloacetate

Quick Test 19

1. Where does the citric acid cycle take place within a cell?
2. In an animal cell, which substance is produced from pyruvate if no oxygen is present?
3. Name the two hydrogen carriers in the citric acid cycle.
4. Which stage of cell respiration involves a chain of protein carrier molecules?
5. Describe the role of ATP synthase in the production of ATP molecules.
6. Name the final hydrogen acceptor in the electron transport chain.
7. Apart from glucose, name two other respiratory substrates.

Metabolic rate

The **metabolic rate** of a living organism refers to the speed at which glucose is broken down into energy (ATP), carbon dioxide and water during cell respiration.

Comparison of metabolic rates between different organisms

The metabolic rates of different living organisms can be compared by measuring one of the following:

1. Oxygen consumption

2. Carbon dioxide output

3. Energy output in the form of heat.

The oxygen consumption per hour of a small organism can be measured using a simple **respirometer**.

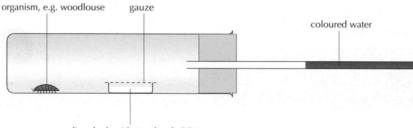

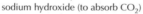

Carbon dioxide produced by the living organism is absorbed by the chemical. Oxygen consumed lowers the air pressure in the apparatus and can be quantified by measuring how far along the capillary tube the coloured water has travelled, taking the place of the consumed oxygen.

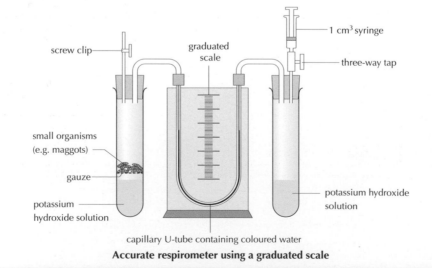

Accurate respirometer using a graduated scale

Complex organisms that have a high metabolic rate have a greater demand for oxygen. These organisms have evolved efficient cardiovascular transport systems to deliver oxygen from the lungs to body cells.

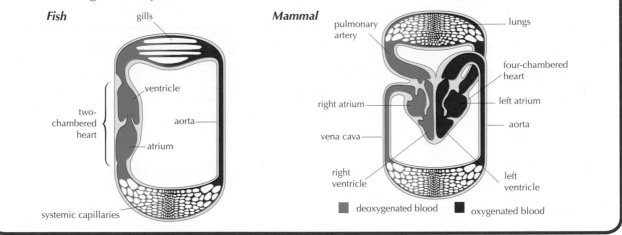

Transport systems of animal groups

Animal group	Number of heart chambers	Type of circulation	Gas exchange system
Fish	2 1 atrium, 1 ventricle	Single – blood passes through heart once	Gills
Amphibians	3 2 atria, 1 ventricle	Double – blood passes through heart twice	Damp skin and mouth mainly Lungs used when needed
Reptiles	3 2 atria, 1 ventricle	Double	Lungs containing alveoli
Birds	4 2 atria, 2 ventricles	Double	Lungs containing parabronchi
Mammals	4 2 atria, 2 ventricles	Double	Lungs containing alveoli

Physiological adaptations for low oxygen niches

The niche of an organism is the role it plays within the environment, defined by, for example, the type of food consumed, predators, temperature and oxygen availability.

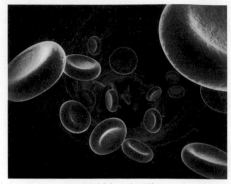

Red blood cells

1. High altitude niche – Human populations living at high altitude are exposed to a much lower concentration of atmospheric oxygen than those living at sea level. These populations have responded to the low oxygen environment by producing a greater number of red blood cells. This improves the efficiency of oxygen delivery to the cells of the body.

2. Deep water niche – Diving mammals, such as seals, conserve oxygen breathed in at the surface by slowing down their heart rate and metabolic rate. They must return to the surface following a dive to gain more oxygen by breathing air.

Diving seal

VO$_2$ max

'VO$_2$ max' is the maximum volume of oxygen that a person is able to breathe in during strenuous exercise. This is used to measure the efficiency of the cardiovascular system. The higher the VO$_2$ max value, the more efficient the metabolism and cardiovascular system.

Quick Test 20

1.	What is meant by the term metabolic rate?
2.	Which three factors in a respiring organism can be measured in order to calculate metabolic rate?
3.	What is the name of the apparatus used to do this?
4.	Which feature of the lungs in birds makes them efficient at transporting oxygen to the body?
5.	How many heart chambers are found in both reptiles and amphibians?
6.	Describe one physiological adaptation of a named organism living in a low oxygen environment.

Conformers and regulators

Conformers

The **internal environment** of an organism is made up of body cells and their surrounding tissue fluid, together with temperature, water content and glucose content.

Conformers **cannot regulate** their internal environment and therefore cannot control their metabolic rate.

The body temperature of a conformer depends upon the temperature of the external environment.

Conformers include reptiles and invertebrates.

Advantage

As conformers have no physiological mechanism for controlling metabolic rate, energy costs to the organism are low.

Disadvantage

Conformers can only occupy a narrow range of ecological niches and are less able to adapt to and survive any changes in the external environment.

Regulators

In regulators, such as mammals and birds, metabolism is used to control the internal environment using physiological mechanisms. The internal environment remains stable despite changes in the external environment. This is called **homeostasis**.

TOP TIP

Remember the 3Cs: conformers **c**annot **c**ontrol their internal environment by physiological means.

Advantage

Regulators can occupy a wide range of ecological niches.

Disadvantage

Energy costs to the organism are high due to the physiological mechanisms needed to keep the internal environment stable.

Temperature control in regulators

Negative feedback mechanism

Mammals are regulators and are able to maintain an internal body temperature of 37 degrees to facilitate optimum enzyme activity and movement of molecules by diffusion using a system of negative feedback control.

- A change in temperature of the internal environment is detected by **receptor cells**.
- A nervous signal is sent from receptor cells to the **effector organ**, the skin.
- The effector organ brings about changes to restore the body temperature to the **normal or set point**.

	Increase in body temperature	Decrease in body temperature
Receptor cells	Receptor cells in hypothalamus monitor temperature of blood Receptor cells in skin monitor external temperature	Receptor cells in hypothalamus monitor temperature of blood Receptor cells in skin monitor external temperature
Type of signal sent from receptor cells	Nervous	Nervous
Response of effector organ (the skin)	• Sweat glands activated • Hairs on skin lie flat • Vasodilation – blood vessels widen, bringing blood to surface of skin to lose heat by radiation	• Sweat glands stop producing sweat • Hairs on skin stand up to trap air as insulator • Vasoconstriction – blood vessels narrow, diverting blood away from surface of skin

TOP TIP

Giving alcohol to a person who is very cold causes vasodilation, increasing heat loss from the body which can lead to hypothermia.

TOP TIP

Regulators use physiological mechanisms to control their internal environment. These include:
- Control of blood glucose.
- Control of blood water.
- Control of body temperature.

TOP TIP

Metabolic rate decreases when body temperature increases, and vice versa.

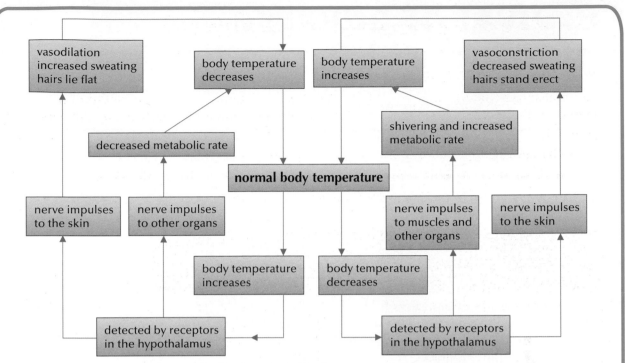

Negative feedback control of body temperature in mammals.

A human is able to regulate body temperature for optimum enzyme function.

Quick Test 21

1. What is the advantage to conformers of not being able to regulate their internal environment?
2. Explain why conformers can only occupy a very narrow range of environments.
3. Give two examples of physiological mechanisms used by regulators to control their internal environment.
4. What is the disadvantage of physiological control to regulators?
5. Describe the principle of negative feedback control.
6. Where are the receptor cells located in the control of body temperature?
7. Name the effector organ in the control of body temperature.

Survival of adverse environmental conditions

Some organisms can survive major changes in environmental conditions such as temperature and drought, through **dormancy** or **migration**.

Dormancy

Dormancy is a stage in the lifecycle of an organism where metabolic rate decreases and growth of the organism stops.

Types of dormancy

1. **Predictive** – Organisms enter into a dormant stage **before** environmental conditions deteriorate. For example, bears hibernate before the onset of winter.

2. **Consequential** – Organisms enter into a dormant stage **after** a significant detrimental change in environmental conditions.

3. **Aestivation** – A state of inactivity, during periods of high temperature or drought, occurring in some animals such as lungfish, where metabolic processes are slowed down.

4. **Daily torpor** – Some species of small birds and mammals with high metabolic rates reduce their metabolic rate every 24 hours by becoming inactive. This helps to conserve energy.

Lungfish enter a period of aestivation during periods of drought

Migration

Metabolism is sustained in some animals by relocating to another environment when food is in short supply or temperature falls.

This is called migration and avoids **metabolic adversity**.

For example, swallows migrate from Africa to Britain in summer when insects are plentiful, reducing competition for available food. They return to Africa in winter when there are no flying insects available as a food source.

Migrating behaviour in animals is a result of both **innate** and **learned behaviour**.

Innate behaviour patterns are inherited, whereas learned behaviour occurs as a result of trial and error experiences.

Extremophiles

Prokaryotes such as bacteria, which are able to tolerate extremes of temperature, pressure, salinity and pH within their environment, are called extremophiles.

They can thrive in conditions which would be lethal to other living organisms.

This is as a result of the evolution of enzymes within their cells which are tolerant of extreme conditions.

> **TOP TIP**
>
> Remember PCR from Unit 1? Heat tolerant DNA polymerase is extracted from a **thermophile**, a species of bacteria which lives inside the rim of volcanoes! This enzyme is not denatured by temperatures of 70–80 degrees.

Methanogens

Methanogens are bacteria which live in anaerobic (without oxygen) conditions, and generate methane gas from carbon dioxide and hydrogen. They are often found in swamps living in stagnant water.

Sulfur bacteria

Sulfur bacteria are found living in sulfur springs which are deadly to other living organisms, and use hydrogen sulfide gas in photosynthesis instead of water.

Yellow sulfur bacteria growing on the edge of a volcanic sulfur spring

Quick Test 22

1. In which type of dormancy does an animal enter into a dormant stage following the onset of winter?
2. What is the advantage of daily torpor to small birds and mammals which have a high metabolic rate?
3. Which type of behaviour in animals avoids metabolic adversity?
4. Explain the difference between innate and learned behaviour.
5. Give two examples of environments inhabited by extremophiles.
6. How are sulfur bacteria able to photosynthesise in sulfur springs?

Environmental control of metabolism in prokaryotes

Growing cells

Metabolism of algae, fungal, archaea and bacterial cells can be controlled during culture in the laboratory by changing environmental conditions in order to increase the final level of product.

Culturing conditions which can be altered:

1. Glucose concentration
2. Temperature
3. pH levels using buffer solutions
4. Oxygen concentration by aeration of growth medium.

Micro-organisms must be grown in **aseptic** conditions, to eliminate the effects of wild micro-organisms in the air which may contaminate the culture medium.

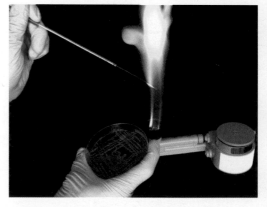

Using a Bunsen burner to sterilise

Micro-organisms must be provided with a respiratory substrate in the growth medium, together with 'building blocks' for the **biosynthesis** of new molecules such as a supply of amino acids and DNA nucleotides.

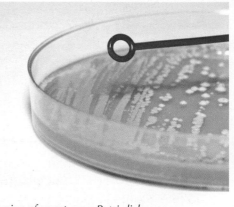

Colonies of yeast on a Petri dish

Additional compounds such as fatty acids and specific vitamins must be added to the growth medium of certain complex micro-organisms in order to culture them in the laboratory successfully.

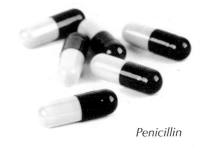

Penicillin

Alcohal

Phases of growth of cell cultures

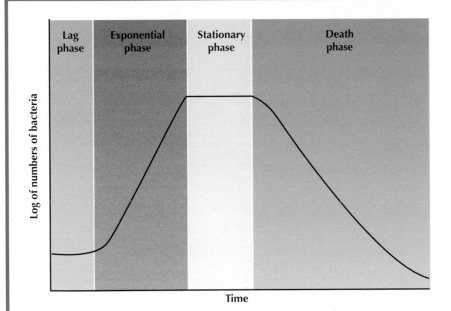

Time

Generation time (doubling) is the time taken for one cell to divide by mitosis into two daughter cells.

- Lag phase – Bacteria induce enzymes to digest culture medium, there is little increase in the cell number.
- Exponential (log) phase – The cells divide at the maximum rate.
- Stationary phase – The cells run out of nutrients. Toxic secondary metabolites accumulate, limiting further cell division. The number of new cells produced is equal to the number of cells killed by the secondary metabolites. There is no growth.
- Death phase – The number of cells killed is greater than the number of new cells produced. The population begins to die out.

Control of metabolism

The metabolism of prokaryote cells grown in culture can be controlled through the addition of metabolic precursors, inducers or inhibitors to produce the desired end product.

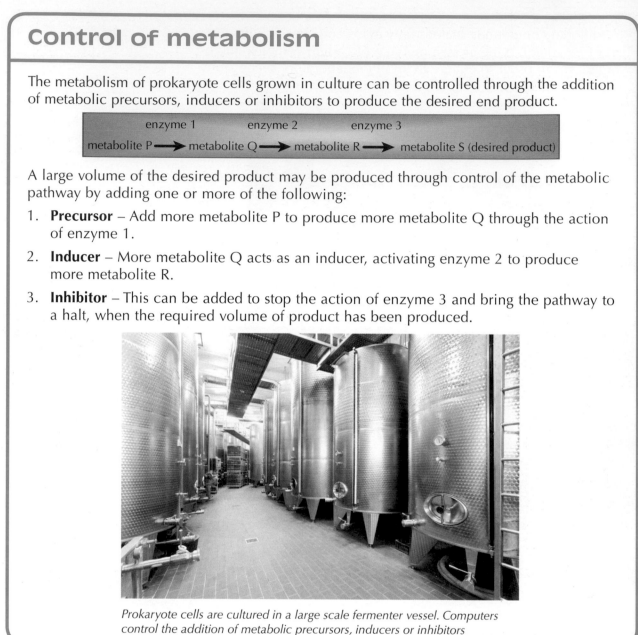

enzyme 1   enzyme 2   enzyme 3

metabolite P ➡ metabolite Q ➡ metabolite R ➡ metabolite S (desired product)

A large volume of the desired product may be produced through control of the metabolic pathway by adding one or more of the following:

1. **Precursor** – Add more metabolite P to produce more metabolite Q through the action of enzyme 1.

2. **Inducer** – More metabolite Q acts as an inducer, activating enzyme 2 to produce more metabolite R.

3. **Inhibitor** – This can be added to stop the action of enzyme 3 and bring the pathway to a halt, when the required volume of product has been produced.

Prokaryote cells are cultured in a large scale fermenter vessel. Computers control the addition of metabolic precursors, inducers or inhibitors

Quick Test 23

1.	Which cell culture conditions can be altered when growing micro-organisms in a laboratory to maximise the output of a final useful secondary metabolite?
2.	Identify two molecules which must be present in the growth medium to allow biosynthesis to take place.
3.	What is meant by the term 'generation time' in cell culture?
4.	During the growth of a cell culture, what occurs during the 'exponential phase'?
5.	In which phase of growth do cultured cells produce secondary metabolites?
6.	Give two examples of secondary metabolites which are useful to humans.
7.	What is the function of a precursor in a metabolic pathway?

Genetic control of metabolism

Wild types of micro-organism

Wild types of micro-organism can be selected for desirable genetic traits, such as the ability to produce large quantities of secondary metabolite and the ability to grow on an inexpensive growth medium.

Wild types of micro-organism can also be improved by:

* Selective breeding – Only those cells with desirable traits are cultured.
* Mutagenesis – Cultured cells are exposed to mutagenic agents such as ultraviolet light to induce desirable mutations.
* Genetic engineering – A useful gene from another organism is introduced into bacterial DNA, allowing the bacteria to code for and produce a new protein.

TOP TIP

Bacterial cells have two circular strands of DNA. The smaller circular strand is called a plasmid and is used in genetic engineering.

Horizontal transfer of genetic material

Some strains of bacteria can transfer chromosomal DNA (plasmids) between cells to produce new strains of bacteria that may have desirable characteristics. This is called **horizontal transfer**.

1. Bacterial cell forms o conjugation tube
2. Conjugation tube joins to neighboring cell
3. DNA plasmid passes into neighboring cell
4. Both bacterial cells contain a similar set of genes contained in the plasmid

conjugation tube

plasmid

Sexual reproduction in fungi and yeast

New improved strains of fungi and yeast can arise through **meiosis** during sexual reproduction, and the merging of genotypes.

Recombinant DNA technology

Also referred to as 'genetic engineering', this is the process of transferring a desirable gene from one species to another. In addition, genes may be introduced that prevent the survival of micro-organisms outside the culture vessel.

Process of gene transfer

- A plasmid in a bacterial cell is cut open using the enzyme **restriction endonuclease** which cuts at a specific base sequence called a **restriction site**.
- Restriction endonuclease cuts out the desired gene from the DNA of another organism.
- The desired gene is inserted into the open bacterial plasmid and sealed in place by the enzyme **DNA ligase**.
- The bacterial plasmid containing the new gene is called a **recombinant plasmid**.

Plasmid is treated with restriction enzymes to cut it open. The open ends are 'sticky'.

Meanwhile a fragment containig the gene that is to be inserted into the plant is cut out of the plant genome, also using restriction enzymes.

sticky end

gene to be inserted

sticky end

sticky end

The fragment becomes incorporated into the plasmid, forming a recombinant plasmid. Enzyme DNA ligase seals transferred gene into recombinant plasmid.

recombinant plasmid

The recombinant plasmid is then inserted back into the bacterium, producing a recombimant bacterium.

Bacteria are used to infect the plant to be modified. Usually a bacterial species such as *Agrobacterium* is chosen. This soil bacterium commonly lives inside plant roots.

plant DNA

Infection of the plant allows the DNA fragment to become incorporated into the genome of the plant. The new gene is expressed and the plant has been genetically modified.

An organism that carries a gene transferred from a different organism is called **transgenic**.

Origin of replication

'Origin of replication' refers to a specific set of genes within a plasmid that control replication of the plasmid during cell growth and expression of the transferred gene, resulting in the generation of the required product, or secondary metabolite.

Marker gene

A 'marker gene' is a gene that acts as a recognisable 'tag' for another closely linked gene.

A specific marker gene within a recombinant plasmid can be detected, identifying which bacterial cells contain the transferred gene following cell division and growth.

The marker gene and transferred gene travel together in a plasmid.

Artificial chromosomes

Artificial chromosomes are constructed in the laboratory using free DNA nucleotides, and are able to carry a much larger gene into a host cell.

Both artificial chromosomes and plasmids act as **vectors**, transporting a gene from the DNA of one species to the DNA of another species.

> **TOP TIP**
>
> Remember:
> - DNA restriction endonuclease acts as 'scissors', cutting out a gene.
> - DNA ligase acts as 'glue', sealing the transferred gene into place.

Issues with the use of micro-organisms

1. **Hazards** – Working with micro-organisms in the biotechnology industry can be hazardous for the workforce. Some micro-organisms may have the potential to cause a harmful reaction in humans. Risk assessments must be carried out before production of a product, such as a new drug, or production of a genetically modified micro-organism begins.

2. **Ethics** – The decision as to which drug is produced using biotechnology is determined largely by financial returns on sales, rather than on who may benefit. Biotechnology companies keep their processes and discoveries highly secret for economic reasons, and do not share their knowledge with the wider scientific community.

Quick Test 24

1. Which desirable genetic traits could be selected for, through the selective breeding of wild strains of micro-organism?

2. Other than selective breeding, identify two ways in which wild strains of micro-organism could be genetically improved.

3. Describe the process of horizontal transfer of genetic material in prokaryotes.

4. Which part of the genetic material of a bacterial cell is used in recombinant DNA technology?

5. Explain the role of a restriction endonuclease enzyme in the process of gene transfer.

6. What is the function of a marker gene?

Food security and sustainable food production

Food security

Food supply is the provision and distribution of food to a consumer.

Food security is defined by the World Health Organisation (1996) as being 'when all people at all times have access to sufficient, safe and nutritious food to maintain an active and healthy life'.

Food security depends upon:

1. Food availability – market supply
2. Food access – having enough money to buy or grow food
3. Food use – healthy eating based upon knowledge of nutrition.

Food availability depends upon efficient, sustainable agricultural methods of producing food crops.

Crop production depends upon the ability of soil to support plant growth, together with environmental factors such as temperature and water availability.

In Third World countries, where there is poor soil and long periods of drought, people have very little food security.

Producing crops in a limited area

Production of food crops such as wheat, maize, potatoes and rice grown on a small area of land can be increased by:

* growing a genetically high yielding species of crop plant or **cultivar**
* applying nitrogen based fertilisers
* reducing pests and diseases which attack the crop
* reducing competition between growing crop plants.

Plant growth and productivity

Plants are able to convert light energy into chemical energy in the form of carbohydrate (glucose), through the process of **photosynthesis**. Glucose is stored in plant cells as starch. When water is removed from plant cells, the more glucose that has been stored as starch during photosynthesis, the greater the **dry mass** of the cells in grams.

Net assimilation is the increase in dry mass due to the manufacture of carbohydrate in the plant cell **minus** the carbohydrate (glucose) used up in cell respiration to produce ATP.

Net assimilation can be calculated by measuring the increase in dry mass per square metre of crop leaf area.

Only 10% of this energy is passed on at each stage or **trophic level** of a food chain and is available for growth, which is relatively inefficient. Cattle, sheep and pigs are less productive than plants due to energy loss at the first trophic level and inefficient conversion of plant material into saleable meat.

Primary productivity of a growing crop is the rate at which light energy is converted into carbohydrate during photosynthesis within a period of time. In global ecosystems, this can be calculated as kilocals/m²/year.

Global ecosystems have different levels of primary productivity.

An increase in productivity results in an increase of **biomass**, or total crop yield.

Economic yield is that biomass from the crop which can be sold, for example, rice grains.

Harvest index is calculated by dividing the dry mass of the economic yield by the dry mass of the biological yield (whole crop). The higher the harvest index value, the higher the economic value of the crop.

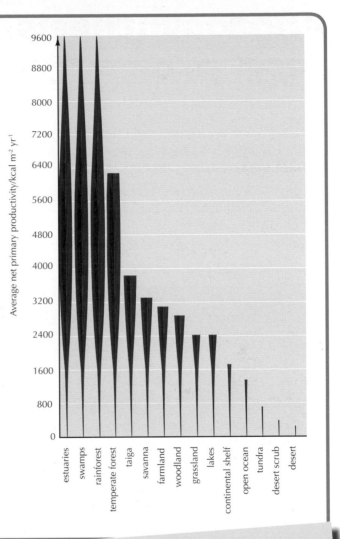

TOP TIP

Productivity is affected by:
- light intensity
- abiotic factors such as temperature, water availability
- soil type
- efficiency of leaves at capturing available light – leaf shape, angle, area
- crop density
- availability of nutrients.

Quick Test 25

1.	Define the term 'food security'.
2.	Identify three factors upon which food security depends.
3.	Suggest two ways in which the yield of a crop, grown on a small area of land, may be increased.
4.	What is meant by the term 'primary productivity'?
5.	Explain what is meant by 'net assimilation' in a growing crop.

Photosynthesis

Photosynthetic pigments

'Photo' means light and 'synthesis' means 'to make'.

Photosynthesis is the way in which plants use light energy to make carbohydrate in the form of glucose.

Plants contain coloured chemical compounds called **pigments**. These photosynthetic pigments are found in the **granum** of the chloroplast. Each pigment absorbs different colours or wavelengths of light, which make up the visible spectrum.

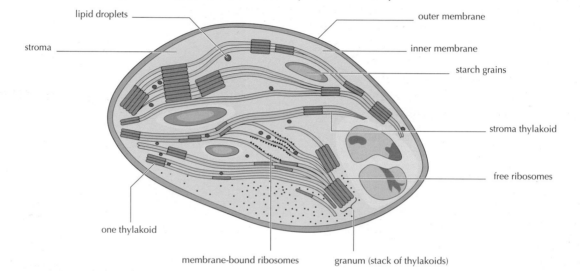

There are four main photosynthetic pigments in a green plant:

1. chlorophyll a
2. chlorophyll b
3. xanthophyll (carotenoid)
4. carotene (carotenoid).

Pigments can be extracted from a green plant and identified using the process of thin layer chromatography to produce a chromatogram. The pigments have different solubilities in the solvent. Those that are most soluble travel furthest, those that are least soluble are found nearer the origin of the chromatogram.

Chlorophyll a and b absorb most light energy at the **blue** and **red** end of the spectrum.

The carotenoids, xanthophyll and carotene, extend the absorption spectrum by absorbing light at the **yellow** and **orange** regions of the spectrum. This energy is then passed on to chlorophyll a.

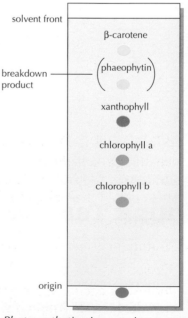

Photosynthetic pigments in a green plant by chromatography

Absorption spectrum

An absorption spectrum graph shows which colours (wavelengths) of light are absorbed by each of the four photosynthetic pigments. Plants absorb most light at the blue and red end of the spectrum. Green light is not absorbed but is reflected to the human eye, which is why many plants appear to be green.

Action spectrum

An action spectrum graph shows how good each colour of absorbed light is at 'driving' photosynthesis. The rate of photosynthesis is highest in blue and red light, the ends of the spectrum where most light is absorbed by green plants. There is a little photosynthetic activity in the yellow and orange parts of the spectrum due to some absorption of these wavelengths of light by the carotenoid pigments.

TOP TIP

Remember that the **'action'** referred to in the action spectrum is **photosynthesis**!

TOP TIP

Not all light landing on a leaf is absorbed into the cells: Some light passes straight through the leaf – **transmitted**. Some light bounces off the surface of the leaf – **reflected**.

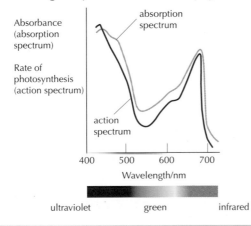

Absorbance (absorption spectrum)

Rate of photosynthesis (action spectrum)

absorption spectrum

action spectrum

400   500   600   700

Wavelength/nm

ultraviolet   green   infrared

Process of photosynthesis

The manufacture of glucose in plant cells using the raw materials of water and carbon dioxide gas together with light energy, occurs in two stages. The first stage called **photolysis** takes place in the granum of the chloroplast within a plant cell. The second stage called the **Calvin cycle** takes place in the **stroma** of the chloroplast.

Photolysis

Photolysis is light dependent.

'Photo' means light and 'lysis' means to break.

Photolysis occurs in the granum of the chloroplast and uses light energy to break the chemical bonds within a water molecule, releasing the hydrogen and oxygen atoms.

Stages of photolysis

- Absorbed light energy causes electrons within pigment molecules to reach a 'high energy' state.
- Some high energy electrons flow along an electron transport chain, rotating the enzyme molecule ATP synthase, located on the plasma membrane of the granum, generating a molecule of ATP from ADP + Pi.
- Other high energy electrons break the chemical bonds between hydrogen and oxygen atoms in a molecule of water.
- Oxygen and hydrogen are released.
- Oxygen moves though the plant cells by diffusion and exits the leaf through the stomata.
- Hydrogen is picked up by the co-enzyme **NADP**, to form NADPH.
- Both NADPH and ATP produced during photolysis are essential to the next stage of photosynthesis, called the **Calvin cycle**.

Diagram of photolysis

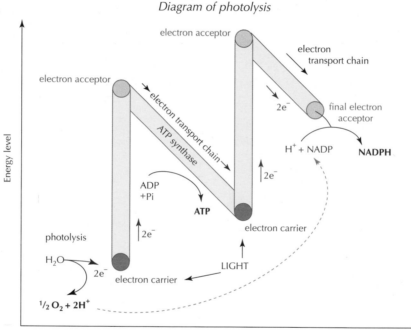

Calvin cycle

Calvin cycle is light independent.

The Calvin cycle occurs in the stroma of the chloroplast, and is dependent upon NADPH and ATP being produced during photolysis.

Stages of the Calvin cycle

- Carbon dioxide enters the leaf through the stomata, and moves by diffusion into the chloroplast.
- Carbon dioxide combines with **RuBP** (ribulose biphosphate) catalysed by the enzyme **RuBisCo** (ribulose biphosphate carboxylase oxygenase!).
- **3-phosphoglycerate** is produced and, using the ATP and NADH from photolysis, is converted into **glyceraldehyde-3-phosphate (G3P)**.

> ## TOP TIP
>
> The co-enzyme that picks up hydrogen in cell respiration is NAD, and in photosynthesis it is NAD**P** – remember **P** for photosynthesis!

- G3P has two possible fates; some is used to make more RuBP, which is continually being used up, and some is used to make glucose.
- Glucose can then be used in cell respiration, stored as starch grains in the chloroplast or polymerised to make cellulose for new cell walls.

Quick Test 26

1. Name the four photosynthetic pigments found in green plants, and underline the carotenoids.
2. Explain the difference between an absorption and action spectrum.
3. Where does photolysis take place?
4. Which two molecules produced during photolysis are essential for the Calvin cycle?
5. Name the enzyme which catalyses the reaction between carbon dioxide gas and RuBP.
6. State the two possible fates of glyceride-3-phosphate produced in the Calvin cycle.
7. Name the hydrogen carrier in photosynthesis.

Plant and animal breeding

Productivity of plants and animals in agriculture in order to provide a sustainable source of food can be improved through genetics.

Field trials

Different species of crops, such as grass, cereals and potatoes, can be grown in small replicate plots. Each set of plots are exposed to different treatments, then harvested in order to calculate the economic yield. This information allows farmers to select a species that will give the highest yield within local growing conditions.

Field trials can also be used to generate data on the performance of genetically modified crops, or new cultivars.

Genetic selection

Animals and plants with desirable genetic characteristics are selected for breeding, ensuring these genetic characteristics are passed on to the next generation.

Outbreeding involves fertilisation of gametes from two **unrelated** members of the **same** species. Animals and plants are naturally outbreeding.

Intensive genetic selection can lead to **inbreeding**, involving the fertilisation of gametes from closely related individuals within a species. This can be used to produce individuals with the required desirable genes after several generations.

This can also lead to inbreeding depression, where numerous recessive homozygous alleles appear in the phenotype of the offspring.

For example, using the same ram on a flock of ewes for more than two years can lead to lambs being born with a split lower eyelid which is a sign of inbreeding depression.

Self-pollinating plants are naturally inbreeding, where harmful recessive alleles are removed by natural selection.

> **TOP TIP**
> - **Ph**enotype – **ph**ysical appearance of offspring.
> - Genotype – alleles (forms of a gene) carried by offspring on DNA.
> - Dominant alleles (strong) – show up in phenotype.
> - Recessive alleles (weak) – do not show up in phenotype unless there are no dominant alleles present to mask them.

> **TOP TIP**
> Possible desirable genetic characteristics:
> - **Plants** – disease resistance, larger grains or fruits, drought resistance.
> - **Animals** – high conversion of food to body mass, high fertility rate, disease resistance.

Cross breeding animals

Crossing individuals of different breeds within a species can result in both genetically and physically stronger F1 offspring, and produce desirable characteristics from each parent. This is called **hybrid vigour**.

Scottish Half Bred sheep are a cross between Cheviot and Border Leicester breeds. The Half Bred has a large body frame inherited from the Border Leicester and high muscle mass inherited from the Cheviot.

Cheviot

Border Leicester

Back cross (test cross)

Back crossing is required to ensure the proliferation of desirable genes from the F2 generation onwards. This is a way of identifying those individuals that are homozygous for desirable genetic traits, which can be used for breeding. A **back cross** can also identify those individual animals carrying undesirable heterozygous recessive alleles which cannot be detected in their physical appearance, but will be revealed within the F1 offspring.

For example, Animal 1, which may be homozygous or heterozygous for a desirable genetic trait, is crossed with a known homozygous recessive Animal 2 for that same trait. The F1 phenotypes are examined. If Animal 1 is homozygous, the desired trait will show up in all offspring, and Animal 1 is kept for breeding to ensure the trait is passed on to the F3 generation and so on.

If Animal 1 is heterozygous for the desired trait, it will appear in only half of the offspring, and Animal 1 will be rejected for further breeding.

Aberdeen Angus cattle have a desirable dominant allele for black coat (B), and a less desirable recessive allele for red coat (b). An animal with a black coat could be true breeding (BB), or heterozygous (Bb).

To find out whether the animal is heterozygous or true breeding, it is crossed with another true breeding (homozygous) recessive animal (bb), which has a red coat. This is called a **back cross** or **test cross**.

Result:

- If calves from cross **all** have black coats, they are all Bb and the black coated parent must be BB or true breeding.
- If some calves from cross have black coats (Bb) and some have red coats (bb), the black coated parent must be heterozygous (Bb).

TOP TIP

Genetically true breeding animals command higher market prices when sold as breeding stock.

Genome sequencing

By identifying the base sequences of an organism's genome using PCR and **bioinformatics**, desirable gene sequences can be identified directing the management of breeding programmes.

Transgenic animals and plants

When a gene from the DNA of one species is inserted into the DNA of a different species through genetic engineering, the recipient organism is called **transgenic**, and the process is referred to as **genetic transformation**.

A flock of Dutch milk sheep have each had the human gene that codes for the Factor 8 blood clotting protein inserted into their DNA. The sheep are milked twice a day and the human Factor 8 protein in the milk is extracted. This is then used to treat patients suffering from haemophilia, who lack this essential blood clotting protein. These sheep are transgenic or 'genetically modified'.

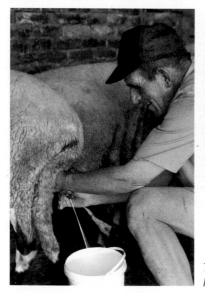

Transgenic tomatoes contain a gene from Atlantic salmon DNA which gives the fruit a deep, uniform red colour.

Transgenic sheep produce milk containing human Factor 8 protein.

Quick Test 27

1.	How can field trials provide a useful way of selecting desirable genetic characteristics in plants?
2.	What is meant by the term 'inbreeding depression'?
3.	Why is crossbreeding animals of different breed within the same species an advantage?
4.	How can an animal or plant which is heterozygous for a genetic trait that does not appear in the phenotype be identified?
5.	Explain the term 'genetic transformation'.
6.	Which name is given to an organism that has received a gene from another species?

Crop protection

Weeds, pests and diseases

Productivity of a crop can be greatly reduced by competition with populations of:

- weeds
- pests and diseases.

Weeds

Annual weeds	Perennial weeds
• Sexual reproduction • Short lifecycle (one year) • Rapid growth • High numbers of seeds produced • Dormant seeds remain viable for long period of time • Examples: goose grass, knotweed, shepherd's purse.	• Asexual reproduction • Long lifecycle (two years+) • Broken pieces of plant can root and grow into new individual plants • Storage organs provide food for plant in autumn/winter • Examples: couch grass, nettles, buttercup.

Pests and diseases

Pests	Diseases
• Insects – cause leaf damage, such as greenfly, blackfly, leatherjackets • Nematodes – round worms in soil attack roots and storage organs • Molluscs – slugs and snails damage leaves	• Fungi – yellow rust on leaves of cereal plants; brown rot on stone fruits, peaches and plums • Bacteria – affect stems, roots and leaves and cause leaf spots, blight and galls • Virus – yellow mosaic of lettuce leaves caused by the lettuce mosaic virus reduces photosynthesis

TOP TIP

A 'weed' is a plant growing in the wrong place!

TOP TIP

Weeds within a growing crop can:
- compete with crop plants for nutrients
- contaminate crop at harvest time
- release chemicals into soil which inhibit growth of crop plant
- provide habitats for pest species.

Control of weeds, pests and diseases

Cultivation

Cultivation and good management of soil help to prevent the build up of weeds, pests and diseases.

1. **Ploughing** – This destroys perennial root systems in the soil and cleans the field of any previous crop residue which may harbour some pests and plant diseases.
2. **Autumn sowing** – Seeds planted in the autumn allow a crop to become established before a surge of pests and diseases in the spring, improving crop tolerance.
3. **Weeding** – Growing weeds are removed by cultivation between rows of crop plants.
4. **Crop rotation** – Planting different crops in each field each year helps to break the lifecycle of pest species.

Chemical control

Herbicides, pesticides and fungicides are chemicals that are sprayed onto a growing crop.

Herbicides are chemicals that kill weeds and are grouped according to how they attack a weed.

Contact	Selective	Systemic
Non-selective, kills all green plants on contact	Attacks only broad-leaved plants. Narrow-leaved plants such as grass are unaffected	Herbicide absorbed into transport system of plant (xylem and phloem). Kills all parts of the plant
Example: 'Resolva'	Example: 'Preen'	Example: 'Roundup'

Pesticides and **fungicides** are chemicals that kill pest and fungal species on plants.

	Contact	Systemic
Pesticides	Kills pest on contact with spray	Spray absorbed by plant, which becomes poisonous to pest
Fungicides	Sprayed on crop to **prevent** growth of fungus, following consultation with the disease forecast for the region	Spray absorbed by plant to **prevent** invasion and growth of fungus

Disadvantages of chemical crop protection

- Toxic to wild animals.
- Chemical remains in soil and surrounding environment following harvest of crop.
- Chemical enters food chain and accumulates in the body tissue of animals. The concentration of the chemical increases through the trophic levels of a food chain, with the predator containing the highest concentration of chemical in body tissues such as fat. This is called **biomagnification**.

- Repeated exposure to the same chemical can produce resistant strains of pests and diseases.

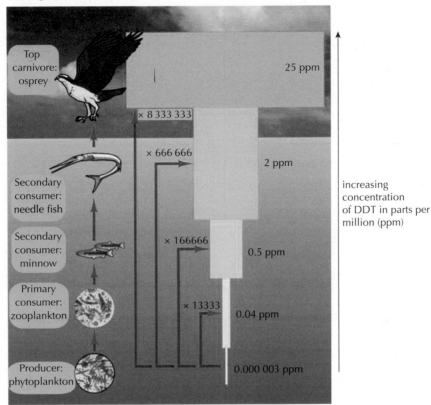

Biomagnification of pesticide DDT in a food chaim

Top carnivore: osprey — 25 ppm

× 8 333 333

Secondary consumer: needle fish

× 666 666 — 2 ppm

Secondary consumer: minnow

× 166666 — 0.5 ppm

Primary consumer: zooplankton

× 13333 — 0.04 ppm

Producer: phytoplankton — 0.000 003 ppm

increasing concentration of DDT in parts per million (ppm)

Biological control

- No chemicals are used.
- Pest population is reduced by the deliberate introduction of a natural predator. For example: a greenfly infestation in a glass house can be biologically controlled by introducing ladybirds which eat greenfly.
- Introduction of a natural parasite or disease may also reduce pest numbers.

Risks of biological control

- Predator, parasite or disease species used to control a pest population could also reduce populations of native species, causing ecosystems to collapse. A predator may feed on more species of prey than the target species, causing imbalance in a food web.

Ladybird attacking aphids on an endangered plant

TOP TIP

Integrated pest management (IPM) uses a combination of chemical and biological techniques to **control** the population size of a pest, instead of killing the whole population.

Examples of biological control

In Australia, a species of moth has been introduced that feeds on the prickly pear cactus. This controls the spread of the prickly pear.

The cane toad was introduced into Australia to control the population growth of cane beetles on which it feeds. However, the came toad itself became a pest as it kills many other small invertebrates. The cane toad has a poisonous skin which will kill any predator that tries to eat it.

Cane toad

Quick Test 28

1. State two factors which may compete with a growing crop, reducing productivity.
2. State three differences between annual and perennial weeds.
3. List three main groups of pest species.
4. Which parts of a plant are specifically targeted by bacterial infections?
5. Describe how cultivation techniques in agriculture may be used to control weeds, pests and diseases in crops.
6. Identify the three main groups of weed killer.
7. What are the disadvantages of: a) chemical control of weeds, pests and diseases, b) biological control methods?

Animal welfare

The care and welfare of animals farmed for food is not only ethically desirable, but has been shown to increase productivity.

Cost	Benefit
Financial investment to improve environmental living conditions for animals	Animals are less stressed, leading to greater productivity and reproductive rate

Indicators of animal stress

- **Stereotypy** – Repetitive behaviour patterns are observed, for example, animal pacing to and fro in a cage.
- **Misdirected behaviour** – A normal behaviour, such as preening feathers in birds, is directed against the animal itself, leading to excessive preening and pulling out of feathers.
- **Failed sexual/parental behaviour** – Animal stress as a result of isolation or living in a poor environment can inhibit reproductive behaviour; young produced under these conditions may be rejected by the parent.
- **Altered levels of activity** – Animals suffering from stress may be recognised by excessive levels of activity in the form of hyper-aggression or excessive sleeping.

Tiger in a cage

Observing and recording behaviour

A 'behaviour' demonstrated by an animal can be defined as 'an observable, measurable response, to either an internal or external stimulus'.

An **ethogram** is a list of all behaviours carried out by an animal during a period of observation.

An ethogram constructed for a domesticated animal in optimal environmental conditions can provide information for farmers regarding those behaviours which indicate an animal is not suffering from stress. This confirms to the farmer that housing and environmental conditions are successfully meeting the needs of the animal.

TOP TIP

Measurement of behaviour:
- **Frequency** – number of times a behaviour occurs during a set period of time
- **Duration** – the time each observed incidence of a specific behaviour lasts
- **Latency** – the time measured between a stimulus being applied and a behavioural response.

A **preference test** is an experiment set up to provide information on which one of two experimental conditions an animal will choose. For example, an animal may be presented with a choice of food or bedding types, or different areas of living space.

Motivation – The incentive to carry out a specific behaviour is called 'motivation'. The strength of motivation may be measured in conjunction with preference testing.

Symbiosis

The co-evolved relationship between two organisms of different species is called a **symbiotic relationship**.

Within a symbiotic relationship, either one or both organisms receives a benefit from the relationship.

Parasitism

In a parasitic relationship, one partner benefits while the other is harmed. The parasite benefits from food and energy provided by the host animal, which in turn loses strength, becomes weak and eventually dies.

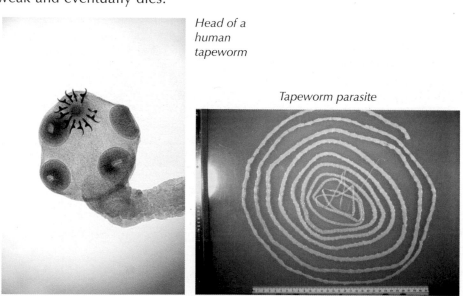

Head of a human tapeworm

Tapeworm parasite

Transmission of parasites

When a host begins to fail, a parasite must move on to a new host. This can be achieved by:

- **Direct contact** – Host animals have physical contact with each other and pass the parasite between them. For example, human head lice.
- **Secondary host** – This is an organism that acts as a temporary, intermediate carrier of the parasite until a suitable **primary host** is found. For example, watercress acts as an intermediate host for the liver fluke parasite in sheep.

- **Vector** – This refers to an organism that transports the parasite from one host to another. For example, the mosquito acts as a vector for the malaria parasite.

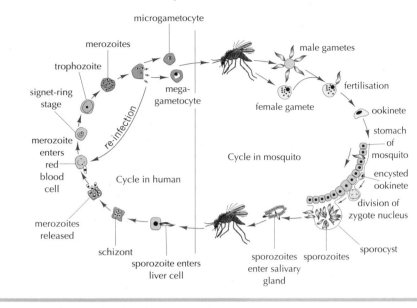

Mutualism

Both species within a mutualistic relationship benefit in terms of food and energy, and are usually compatible in terms of physical structures due to the process of co-evolution.

Mitochondria and chloroplasts

Mitochondria and chloroplasts are cell organelles which began life as free living prokaryotic cells. At some point during the process of evolution they entered the cytoplasm of an anaerobic eukaryotic cell, forming a mutualistic relationship. The prokaryotes benefited from the secure environment inside the eukaryotic cell, which in turn benefited from the oxygen and ATP generated by the chloroplast and mitochondria.

Lichen

Resource distribution

Primates are able to feed on both plants and animals within their own geographical range in a large forest ecosystem. Although mainly herbivorous, chimpanzees will occasionally hunt monkeys cooperatively in a group and share the kill between them. Chimpanzees have also learned from their parents how to use a thin stick to collect termites from inside the trunk of a tree.

Chimpanzee with stick

Taxonomic group

Chimpanzees, baboons and monkeys are all primates, but belong to different classified groups. These are called **taxonomic groups**.

Members of each taxonomic group are similar in terms of resource requirements, habitat and type of social structure, but are able to inhabit different ecological niches.

> **TOP TIP**
>
> **Agonistic** behaviours are seen during situations of conflict, and include fighting and running away.

Quick Test 30

1. What are the advantages of a social hierarchy?
2. How do a group of animals benefit from cooperative hunting?
3. Give one example of a social defence mechanism against predation.
4. Which term is used to describe the behaviour of an individual, which is of survival benefit to others?
5. State one example of an 'ecosystem service' provided by insect colonies.
6. What is the benefit of appeasement behaviours within a group of primates?

Biodiversity

Biodiversity is the term used to describe the millions of different species of animals, plants and micro-organisms on Earth.

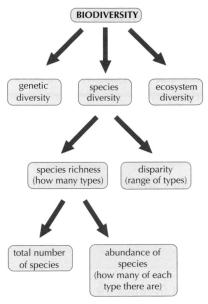

Mass extinction

Mass extinction events are caused by sudden extreme changes in environmental conditions, resulting in the loss or **extinction** of a whole species. Evidence for mass extinction events which have occurred in the past can be found within the fossil record. These events **decrease biodiversity**. Over many generations, biodiversity can begin to increase again due to surviving individuals of a mass extinction event undergoing the process of **speciation**.

Extinction rate

A broad estimate of the extinction rate of a species may be calculated by measuring the number of species that become extinct over a certain period of time. This is easier for large, visible species such as mammals, but very difficult for less visible species such as bacteria, fungi and insects.

Effect of human activities

The extinction of large land animals or **megafuna**, has been primarily as a result of the increased colonisation of land by humans. Many land animals have already become extinct, such as the duckbill platypus and the wooly mammoth.

Glossary

trophic level: a stage in a food chain. Producers are the first trophic level, primary consumers are the second trophic level and so on

vector: an organism such as an insect that passively transports a disease causing bacteria, virus or parasite to another organism

whole genome duplication: a cell receives a double set of chromosomes and becomes polyploid

Answers to Quick Tests

Quick Test 1

1. Adenine pairs with thymine, guanine pairs with cytosine.
2. 5' strand ends with phosphate group.

   3' strand ends with a deoxyribose sugar molecule.
3. One strand has alternate phosphates and sugars, beginning with a phosphate group and ending with a deoxyribose sugar molecule. The opposite strand has alternate phosphates and sugars, beginning with a deoxyribose sugar molecule and ending with a phosphate group. The two strands run in opposite directions and are therefore **antiparallel**.

Quick Test 2

1. Prokaryote has no membrane bound nucleus/circular DNA. Eukaryote has a membrane bound nucleus/linear DNA.
2. A plasmid.
3. Eukaryotic yeast cells have linear chromosomes within the nucleus **and** circular plasmids in the cytoplasm.

Quick Test 3

1. A primer allows the attachment of a chain of DNA nucleotides by the enzyme DNA polymerase.
2. The 3' end.
3. Ligase.
4. Replication involves the formation of many DNA fragments instead of one continuous strand.

Quick Test 4

1. Forensic analysis through DNA profiling at a crime scene. Archaeological analysis of historical plant or animal preserved cells. Paternity tests. Testing for genetic diseases.
2. It is not denatured by high temperatures.
3. A primer is an artificial DNA fragment, used to 'bookmark' either end of the DNA fragment to be copied.
4. 93°C, 55°C, 72°C, (then 93°C).
5. Unique bonding pattern produced by exposing DNA fragments to the process of gel electrophoresis.

Quick Test 5

1. RNA – single strand, ribose sugar, base uracil, smaller than DNA.

   DNA – double strand, deoxyribose sugar, base thymine.
2. Adenine.

Quick Test 14

1. Anabolic, catabolic or integrated.
2. Metabolite 1 may be converted to metabolite 2 and back again. The cell has control over the production and use of metabolite 2.
3. Compartmentalisation isolates and brings together specific metabolites within a metabolic pathway.
4. Phospholipid and protein molecules.
5. Controls entry and exit of molecules to the cell and provides a site for different metabolic pathways.
6. A small structure within the cytoplasm of a cell which has a specific function.
7. Active transport requires energy, and moves molecules from a low concentration to a high concentration.
8. Protein pores contain a channel for the transport of large molecules across the plasma membrane.
9. Active transport and the sodium potassium pump.

Quick Test 15

1. A mutation in a gene would result in the enzyme coded for not being able to function, causing a 'block' at this step of the metabolic pathway.
2. The structural gene coding for the enzyme B-galactosidase is only 'switched on' when lactose is present.
3. Regulator gene/operator gene/structural gene.
4. Structural gene.
5. Regulator gene.
6. Lactose.

Quick Test 16

1. Lowering activation energy speeds up the rate of a chemical reaction.
2. Induced fit describes the ability of the active site of an enzyme to alter shape slightly to accommodate an alternative substrate molecule.
3. The rate of reaction increases until all active sites are full, then it remains constant.
4. Multi-enzyme complexes have more than one function and are able to catalyse much faster reactions, than if working independently.
5. Orientation of substrate molecules within the active site lowers activation energy.

Quick Test 17

1. Inhibitor slows down the rate of reaction.
2. Non-competitive inhibitor.
3. Increase enzyme or substrate concentration.

4. Heavy metals mercury and lead.

5. Non-competitive inhibitor attaches to enzyme.

6. End product inhibition controls production of product molecules within a metabolic pathway.

Quick Test 18

1. Muscle contraction, cell division, replication of DNA, repair, transport, transmission of nerve signals, synthesis of new molecules.

2. ADP + Pi to ATP

3. ATP is constantly being broken down and regenerated, very little is stored by the body.

4. NAD – nicotinamide adenine dinucleotide.

5. Pyruvate.

6. Cytoplasm.

7. 2 ATP needed to break glucose molecule, 4 ATP generated from broken glucose bonds, net gain to cell of 2 ATP.

Quick Test 19

1. Central matrix of mitochondrion.

2. Lactic acid.

3. NAD and FAD.

4. Electron transport chain.

5. ATP synthase is a protein molecule rotated by the flow of electrons, resulting in the phosphorylation of ADP + Pi to ATP.

6. Oxygen.

7. Starch, glycogen, protein, fat.

Quick Test 20

1. The rate at which an organism breaks down glucose during cell respiration into energy (ATP), carbon dioxide and water.

2. Oxygen consumption, carbon dioxide output, energy output as heat.

3. Respirometer.

4. Lungs of birds contain parabronchi, channels in the lung which allow continuous air flow.

5. 3 heart chambers (2 atria and 1 ventricle).

6. Humans living at high altitude produce a higher number of red blood cells than those living at sea level. Seals slow down heart rate and metabolic rate to conserve oxygen during a dive.

Quick Test 21

1. Energy costs to the organisms are low.

2. Conformers are less able to adapt to and survive changes to the external environment.

3. Control of blood water, blood glucose, body temperature.

4. High energy cost to organism.

5. Change detected by receptor cells, send signal to effector organ, brings about changes which restore system to normal.

6. Hypothalamus region of brain, and skin.

7. The skin.

Quick Test 22

1. Consequential dormancy.

2. Reduced metabolic rate conserves energy.

3. Migration.

4. Innate behaviour is inherited, learned behaviour occurs as a result of trial and error experiences.

5. Hot springs, volcanoes, deep water vents, environments without oxygen.

6. Sulfur bacteria use hydrogen sulfide gas instead of water in photosynthesis.

Quick Test 23

1. Glucose concentration, temperature, pH and oxygen concentration.

2. Amino acids and DNA nucleotides.

3. Generation time is the time taken for one cell to divide into two daughter cells.

4. During the exponential phase, cells divide at their maximum rate.

5. Stationary phase.

6. Penicillin, alcohol, medicines.

7. A precursor is a metabolite which is increased in concentration at the beginning of a metabolic pathway, increasing the production of the next metabolite in the pathway.

Quick Test 24

1. Production of large volumes of product (secondary metabolites), able to grow on inexpensive growth medium.

2. Mutagenesis, genetic engineering (recombinant DNA technology).

3. DNA plasmids are transferred between bacterial cells.

4. Plasmid.

5. Restriction endonuclease 'cuts' DNA at a specific base sequence and is used to open a plasmid and cut out a desired gene from the DNA of another species.

6. The presence of a marker gene confirms that a bacterial cell successfully contains a transferred gene within the plasmid.

Quick Test 25

1. Food security – when people have access to good quality, safe food at all times in order to sustain a healthy life.

2. Food availability, access and use.

3. Grow high yielding cultivar, use of nitrogen fertiliser, reduce pests and diseases, reduce competition between plants.

4. 'Primary productivity' is the rate at which light energy is converted to carbohydrate through photosynthesis over a period of time.

5. 'Net assimilation' refers to the mass of carbohydrate produced during photosynthesis **minus** the mass of carbohydrate used up in cell respiration, measured as an increase in dry mass per unit of leaf area.

Quick Test 26

1. Chlorophyll a, chlorophyll b, xanthophyll and carotene.

2. Absorption spectrum – shows the colours of light absorbed by each pigment.

   Action spectrum – shows how good each colour of light is at 'driving' photosynthesis.

3. Granum of chloroplast.

4. ATP and NADPH.

5. RuBisCo.

6. Production of glucose and production of more RuBP.

7. NADP.

Quick Test 27

1. Replicate plots of different plant species grown in a variety of different conditions can be used to provide information on how to improve economic yield.

2. 'Inbreeding depression' occurs when related members of the same species interbreed, and results in undesirable recessive alleles appearing in the phenotypes of the offspring.

3. Crossbreeding animals results in offspring with hybrid vigour.

4. A test cross can be used to identify heterozygous individuals.

5. 'Genetic transformation' refers to the process of inserting a single gene from one species into the DNA of a different species.

6. Transgenic.

Quick Test 28

1. Weeds, pests, diseases.

2. Annual weeds – sexual reproduction, live for one year, rapid growth/lifecycle, produce many seeds. Perennial weeds – asexual reproduction, live for more than two years, food reserves in storage organs, produce small number of seeds.

3. Nematodes, insects and molluscs.

4. Stem, roots and leaves.

5. Crop rotation – breaks pest lifecycle. Ploughing – destroys perennial weeds. Autumn sowing – seedlings establish and grow without competition from spring weeds. Cultivation between rows of crop plants – reduces competition and prevents establishment of weeds.

6. Systemic, contact and selective.

7. (a) Chemical control – chemical residues persist in soil and surrounding environment harming wildlife. (b) Introduced control species – could cause ecosystems to collapse.

Quick Test 29

1. Ethically desirable, increased productivity and rate of reproduction.

2. Stereotypy, misdirected behaviour, failed sexual/parental behaviour, altered levels of activity.

3. Misdirected behaviour.

4. A preference test is set up to give an animal a choice of two conditions. The number of times an animal opts for one of the conditions when the test is repeated is a measure of motivation.

5. Direct contact, secondary host, vector.

6. Mutualism – both organisms benefit. Parasitism – one organism benefits, the other is harmed.

Quick Test 30

1. Reduces physical fighting and the chance of injury. Conserves energy. Genes from strongest individuals passed on to next generation.

2. Each member of a hunting group receives a share of food/larger prey can be hunted.

3. Vigilance behaviour/moving as a herd.

4. Altruistic.

5. Pollination, decomposition, biological control.

6. Reduces fighting and the chance of injury.

Quick Test 31

1. 'Biodiversity' refers to the total number of different species of animal, plant and micro-organism on Earth.

2. Decreases biodiversity.

3. Genetic diversity, species diversity and ecosystem diversity.

4. Species diversity.

5. Exploitation, habitat fragmentation, exotic species, bottleneck effect.

6. Formation of a habitat corridor.

7. An exotic species introduced from another ecosystem which aggressively competes with the native species for resources. The exotic species begins to take over and the native species dies out. Biodiversity is lost.